Transforming Tourism *and* Hospitality

AI-Powered Service Delivery Innovations

Editors

Yashwant Singh Rawal
Amity University Rajasthan, India

Debasish Batabyal
Amity University, Kolkata, India

CRC Press
Taylor & Francis Group
Boca Raton London New York

CRC Press is an imprint of the
Taylor & Francis Group, an **informa** business
A SCIENCE PUBLISHERS BOOK

First edition published 2026
by CRC Press
2385 NW Executive Center Drive, Suite 320, Boca Raton FL 33431

and by CRC Press
4 Park Square, Milton Park, Abingdon, Oxon, OX14 4RN

CRC Press is an imprint of Taylor & Francis Group, LLC

Library of Congress Cataloging-in-Publication Data (applied for)

ISBN: 978-1-032-86651-2 (hbk)
ISBN: 978-1-032-88872-9 (pbk)
ISBN: 978-1-003-54008-3 (ebk)

DOI: 10.1201/9781003540083

Typeset in Times New Roman
by Prime Publishing Services

Preface

The tourism and hospitality industry stands on the cusp of a transformative era, driven by Artificial Intelligence (AI), Internet of Things (IoT), and allied digital innovations. These technologies are no longer peripheral; they have become integral to reshaping customer service paradigms, streamlining operations, and fostering sustainable, secure, and inclusive growth in the sector. This edited volume, *Transforming Tourism and Hospitality: AI-Powered Service Delivery Innovations*, brings together an array of scholarly insights and practical applications that capture the diverse and powerful ways AI is revolutionizing service delivery.

The book opens with a detailed exploration of AI-powered customer service in hospitality, assessing how virtual assistants and chatbots personalize guest experiences and drive satisfaction. From there, the focus shifts to sustainability and AI, especially in reducing food waste and enhancing the sustainable economy, illuminating the critical role AI plays in optimizing resources and supporting environmentally conscious operations.

Subsequent chapters examine smart maintenance practices powered by AI and IoT that enhance asset longevity and operational efficiency. The book further highlights AI's influence on travel, from personalized itineraries and sentiment analysis to fraud detection and dynamic pricing, illustrating how AI enables seamless, secure, and highly customized travel experiences.

Safety and security concerns, a priority in tourism, are addressed through the integration of AI tools such as facial recognition, intelligent navigation, and translation apps. These technologies provide assurance and enhance comfort for travelers navigating unfamiliar environments.

Through robust empirical models and statistical validation methods like Structural Equation Modeling (SEM) and Multiple Linear Regression, several chapters analyze how trust, ethics, transparency, and skepticism influence AI acceptance and customer loyalty. Further discussions delve into the Technology Readiness Index (TRI) and how travelers' technological adaptability affects their pre-, during-, and post-travel experiences.

Technological innovations like Virtual Reality (VR) and smart kitchens offer new dimensions to service excellence, enhancing staff training, customer engagement, and food sustainability. Similarly, blockchain technology is explored as a disruptive force improving transparency, loyalty systems, and operational efficiencies in hospitality.

Real-world case studies, including AI service implementation at JW Marriott Mumbai, virtual airport tours for Indian passengers, and AI-powered itinerary co-creation, provide tangible insights into how digital tools are shaping guest perceptions, reducing anxiety, and fostering trust.

The final contributions underscore how AI-enabled guest service technologies influence consumer behavior, brand loyalty, and overall service experiences. At the intersection of human-AI collaboration, these studies advocate for a strategic blend of efficiency, empathy, and ethics in delivering services that resonate with modern, tech-savvy travelers.

The last chapter, titled **"The Role of Artificial Intelligence in Optimizing Hotel Employee Performance,"** explores how AI-driven automation in hotels impacts employee productivity and service optimization. This study pursues three core objectives: (i) providing an overview of AI in the hospitality industry; (ii) identifying hotel functions enhanced through AI; and (iii) examining the effect of AI-powered HR systems on optimizing employee performance. The findings shed light on the shifting dynamics of hotel workforce management, emphasizing AI's role in enabling staff to focus on complex guest interactions by automating repetitive tasks.

We hope this volume will serve as a comprehensive resource for academics, industry practitioners, policymakers, and students. It not only highlights current innovations but also lays the groundwork for future-oriented research and application of AI in tourism and hospitality service delivery.

Editors

Dr. Yashwant Singh Rawal

Professor, Amity University, Rajasthan, India

Dr. Debasish Batabyal

Associate Professor, Amity University, Kolkata, India

Contents

Chapter 1

AI-Powered Customer Service in Tourism and Hospitality Industry

Malavika Bora*[1] and R. Rajesh[2]

[1]Ph.D. Scholar, Indian Institute of Tourism and Travel Management (IITTM), Noida

[2]Ph.D. Scholar, Utkal University of Culture, Bhubaneswar

INTRODUCTION

Modernizing customer service techniques in the tourism and hospitality industries is rapidly dependent on the use of artificial intelligence (AI), since its application is beneficial in quickly and efficiently examining large volumes of data, addressing some of the most challenging issues plaguing the industry, like timely, customer-specific, and 24/7 service to a diverse client base.

The rise of AI chatbots, virtual assistants, and predictive analytics has made more integrated, data-driven service solutions possible. With the rising customer expectations in the tourism and hospitality industry, there is a growing need to deploy innovative methods for enhancing their experience and boosting engagement (Dang and Nguyen 2023).

Such solutions improve response times and increase the scope of personalization, leading towards a unique customer experience that speaks to individual preferences and needs. AI has become a powerful tool able to take its services even further by improving interactions at all customer touchpoints,

*For Correspondence: malavika.bora98@gmail.com; tourismrajesh@gmail.com

DOI: 10.1201/9781003540083-1

including booking queries, itinerary management, and post-trip customer services. Technological advancement and the need for cost-effectiveness in high-volume spaces have largely driven this transition. The integration of AI has enhanced interactions across various customer touchpoints, ranging from booking inquiries and itinerary management to post-trip assistance (Omind n.d.). This change is primarily fueled by technological progress and the demand for cost-effectiveness in high-traffic industries.

In the tourism and hospitality industry, where users sometimes require support anytime, AI enables businesses to eliminate human boundaries and provide customer support around the clock. This is why the development of AI is expected to continue to grow and change the face of customer service and thereby change business models in tourism and hospitality. Figure 1.1 shows the key drivers of AI adoption in the tourism and hospitality industry.

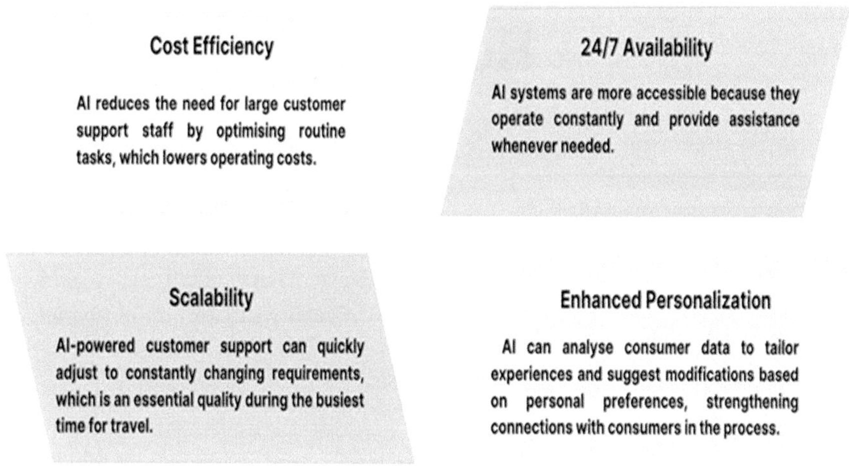

Figure 1.1 Key drivers of AI adoption in the tourism and hospitality industry (*Source:* Author).

UNDERSTANDING CHATBOTS AND VIRTUAL ASSISTANTS

Chatbots and virtual assistants are two of the most significant implementations of AI in the tourism and hospitality industry. These AI-powered applications can sit on messaging apps, websites, or work through voice commands to give instant responses to customers, regardless of the time (Radwan 2024).

Chatbots use artificial intelligence to simulate conversation with users, including text or voice. They serve the purpose of completing what is known as specific goals, like answering frequently asked questions, scheduling appointments, or helping them navigate procedures. Chatbots use either a set of rules or natural language processing (NLP) to work through the questions and provide answers to the users.

Virtual assistants are a step above chatbots in the AI hierarchy. These AI systems are able to do many tasks when integrated with various applications, devices, and services. Most virtual assistants employ sophisticated NLP and machine learning techniques to enable them to comprehend context, personalize responses, and do much more than just answer basic questions. They are able to turn passive, reactive assistants that only answer user queries into proactive assistants that can remind, suggest itineraries, etc., based on the user's preferences. Table 1.1 shows the difference between chatbots and virtual assistants.

Table 1.1 Main differences between chatbots and virtual assistants

Features	**Chatbots**	**Virtual Assistants**
Purpose	Produce information within a limited domain.	Complete a much larger menu of duties.
Main Features	The features are constrained to specific functions, such as responding to common questions, providing product-related information, or navigating a user through a process.	Capability to do multiple tasks such as appointment setting, reminder setting, smart home device control, and personalized recommendations.
Interactions with the Customers	Mainly communicates through text, although some chatbots can also use voice communication.	Utilize voice communication to foster more organic and instinctive conversations.
Adaptability	Can gain insights from interactions to improve responses, typically within a limited range.	Frequently learns and adapts based on user preferences, behaviors, and according to the context of the conversation.
Examples	Customer support chatbots, FAQ chatbots, etc.	Amazon Alexa, Apple Siri, etc.

Source: Author

For example, MAYA is an AI-driven travel planner and itinerary organizer. It doubles as a personalized trip advisor, providing tailored travel plans relative to interests, preferences, and available time. MAYA works less like a robot and more like a virtual assistant, as it interprets user questions and offers personalized recommendations. Although it performs targeted functions such as scheduling, the nature of its design is more intelligent, making it conversational and allowing it to maintain context, thus mimicking virtual assistant traits (Maya—AI assistant for travel companies n.d.).

Chatbots serve an important part in customer service in the tourism and hospitality industries. Rule-based chatbots and AI-based chatbots are the two types. These technologies improve customer communication by providing timely, relevant, and certain assistance.

Rule-based chatbots run on a set of rules and a standard set of flows. Based on set rules and decision trees, the chatbot evaluates user input. They thrive in situations where they need to handle simple repetitive tasks such as responding to frequently asked questions, reviewing changes in reservations, and providing other basic information. Responses can only be triggered via keywords or input patterns. Their syntax only allows them to perform tasks, and they do not possess the ability to adopt algorithms for machine learning to improve or evolve over time.

For instance, **Amtrak**, one of the prominent rail-service providers in the United States, employs 20,000 staff members and caters to 30 million passengers annually (Shah 2023). This agency receives over 35,500 visitors each day and has developed a chatbot named "Ask Julie" to assist its users. Julie enables website visitors to access self-service options and features an upsell capability that allows customers to reserve train tickets and hotel accommodations. The chatbot also supplies route maps for destinations and directs users to the appropriate page for booking rail tickets. Julie facilitates bookings for rail travel, hotels, and car rentals while helping to pre-fill forms, offer guidance, and manage 5 million inquiries each year. The implementation of Julie has significantly boosted the value of bookings for Amtrak.

Since they can apply machine learning (ML), natural language processing (NLP), and sentiment analysis, **AI-powered chatbots** are capable of providing more inventive, more human dialogue in their conversations (Filipsson 2024). They can grasp the context, identify emotions, and adjust their responses accordingly. AI-driven chatbots manage complex inquiries and evolve over time by accumulating more data (Apriorit 2023). They are frequently integrated across various platforms, including websites, social media, and messaging applications.

Table 1.2 Main differences between rule-based and AI-driven chatbots

Feature	Rule-Based Chatbots	AI-Driven Chatbots
Language	Depends on predefined keywords and expressions	Uses natural language processing (NLP) to comprehend and interpret human communication
Generation of Responses	Follows set rules and patterns	Responses based on contextual awareness and learnt patterns
Learning and Adaptation	Does not learn from interactions	Actively accumulates insights and improves through user interactions
Complexity of Tasks	Manages basic and routine tasks	Can handle complex queries and tasks
Flow of Conversation	Rigid, predictable and organized	Organic and dynamic
Development Time	Comparatively quick to develop	Takes more time and resources to train and develop
Maintenance	Needs frequent updates and changes	Requires comparatively less frequent updates and any change or modification
User Experience	Limited and robotic	More human-like experience

Source: Author

AI bots are able to understand the essence of information through NLU while being powered by ML, NLP, and sentiment analysis, enabling them to carry on comparatively convincing and personalized conversations. They can analyze complicated linguistic inputs, learn from prior interactions, and tailor responses. They learn the context, recognize emotions, and respond appropriately. AI-powered chatbots handle complex queries and are improved over time as they learn from accumulated data. They are typically embedded throughout other platforms, from websites to social media to messaging applications.

For instance, KLM's AI chatbot, known as BB (Blue Bot), serves as a leading example of AI-enhanced customer support. Customers of KLM are able to purchase tickets on Messenger with the assistance of artificial intelligence. Acting on behalf of KLM, BB engages with customers on Messenger in a conversational manner to facilitate ticket booking. With this type of artificial intelligence, customers can conveniently secure a ticket without needing a KLM agent's assistance (KLM 2017). Table 1.2 shows the difference between rule-based and AI-driven chatbots.

OVERVIEW OF THE ROLE OF AI IN ENHANCING CUSTOMER EXPERIENCE

Customer experience has to do with the feelings, thoughts, and perceptions customers have at various touchpoints (Chen and Prentice 2024) when engaging with a particular brand or organization, whether online via social media, websites, chatbots, or voice calls, or face-to-face interactions at any physical location or office of a given brand. Organizations regard satisfied customers as repeat or loyal customers, thus wanting to establish trust with them (Barney and Mixon 2024).

AI applies a level of customization and a steeply integrated manner of business that was difficult for typical customer service methods to reach. AI collects user reviews concerning previous tours, including complaint experiences, all automatically generated by the systems when matched against others, just like any regular information site client does on review requests. By looking at AI-generated metrics of customer interest, AI has taken center stage by examining premises based on customer inputs, proclivities, and purchasing history.

Customizations comprise an eminent aspect of AI-enabled customer service in the tourism and hospitality sectors. For instance, an adventurous vacationer may be advised to take outdoor excursions and trekking trails, while affluent travelers are offered ultra-luxurious stays augmented by gourmet dining options. Such personalization is paramount in enhancing customer service with AI in the tourism and hospitality industry.

Below mentioned are the most common parameters to measure the impact of AI on customer service (Qualtrics 2023):

Customer Satisfaction Score (CSAT): Quantifies customer satisfaction with AI-driven communications, as determined by feedback forms or surveys following individual engagements.

Net Promoter Score (NPS): Helps evaluate customer loyalty by ascertaining whether customers would recommend the brand by their experiences.

Average Handling Time (AHT): Assesses the efficiency of AI in resolving customer queries through the average time taken to resolve issues.

First Contact Resolution (FCR): It is a critical metric that evaluates the ability of AI to address customer inquiries promptly and effectively without requiring intervention from human representatives.

These parameters provide a comprehensive structure for assessing the effectiveness of artificial intelligence in enhancing customer interactions at the corporate level, all while maintaining core human service principles and integrating cutting-edge AI advancements.

AI-POWERED CUSTOMIZATION IN CUSTOMER SUPPORT

AI will be the powerhouse of customization since there is a high level of data integration that traditional customer service approaches struggle to mirror. Based on user conversations, preferences, and buying history, AI systems personalize their responses to contact and recommend options that closely mirror individual users' likes and dislikes. For instance, a traveler who prefers adventure tourism might be directed toward suggestions of outdoor activities and hiking trails, while a traveler who prefers luxury tourism would receive suggestions for hotels with high-class amenities and gourmet dining experiences. Personalization is at the heart of AI-based customer service in tourism and hospitality (Bradley 2024).

AI brings personalization to the customer service process that cannot be matched by traditional means. AI will analyze customer interactions, preferences, and historical records and provide unique feedback or suggest choices best for the customer based on this data. For example, a traveler who shows an interest in adventure tourism may receive suggestions for outdoor activities or hiking trails, while a luxury travel enthusiast would receive recommendations for high-end hotels and gourmet restaurants. Personalization is an integral feature of AI-based customer service in the tourism and hospitality sector. Based on data analysis of customer interactions, preferences, and purchases, AI generates highly customized experiences. One example of an AI application in the hospitality sector is Hilton's Connie, an AI-powered robot concierge. Connie, named after Hilton's founder, Conrad Hilton, is the first IBM Watson-enabled robotic concierge, custom-made by IBM to enhance guest experiences, address customer questions, and provide valuable travel tips.

ATTRIBUTES OF PROGRESS IN THE DEVELOPMENT OF AI FOR CUSTOMER SERVICES

Advancements in AI technology in travel customer support are bringing a revolutionary change to the way companies connect with their customers, paving the way for a new world of personalized and efficient assistance (Barten 2024).

This transformation is driven in great part by major developments such as emotion detection and predictive analytics (Takyar 2023).

The technology of emotion recognition uses AI algorithms for analyzing human emotions through various inputs such as voice intonation, text sentiment analysis, and facial expressions. Such capabilities empower AI-driven systems to instantly alter their interactions to provide a sympathetic and context-aware response. Emotion recognition can therefore become invaluable when reacting to customers' complaints or inquiries. For instance, a complaining customer who expresses a tone of frustration over the phone or in chat should be elevated to a qualified human customer service representative. Alternatively, an AI could issue an apology and convey empathy with adaptive language to show understanding and an inclination to resolve the problem.

Predictive analytics employs machine learning algorithms to evaluate past data and project future customer behavior, enabling service providers to forecast customer requirements and deliver proactive solutions. Using large databases of online travel booking forms, search patterns, and specific interests, tourism and hospitality models can be combined, and companies can dive into the market trends. By allowing companies to apply appropriate service strategies, they can devise complete solutions to deliver services effectively and prevent service failures.

CASE STUDY: DIGI YATRA AT THE INDIAN AIRPORTS

India has launched an initiative to make the air travel experience easier and modern, called Digi Yatra, a unique identity program supported by advanced AI that enhances the passenger processing experience at airports (Wong et al. 2021). Started by the Ministry of Civil Aviation, Digi Yatra uses a combination of biometric technology, computer vision, and automation to provide a contactless and seamless flying experience. This move is part of India's larger mission of digital transformation in governance.

Digi Yatra, for the first time in Indian airports, heralds a new era of AI-influenced automation by means of computer vision, facial recognition, and data analytics. AI technologies are harnessed to boost processing efficiency for passenger boarding, as well as to strengthen measures for enhancing safety. Earlier, airports in India relied on ticket checkers and identification verifiers at several checkpoints, causing long waiting times and potential overcrowding during peak travel times. By recognizing the drawbacks of the previous system, the Digi Yatra movement was initiated to automate and better streamline passenger flow from entry to boarding. The platform integrates facial recognition technology with passenger data, enabling real-time validation without the necessity for physical documents. Digi Yatra is indicative of a worldwide trend where airports are implementing smart technology to enhance operational efficiency, reduce wait times, and improve security measures.

Advantages of Digi Yatra are:

1. **Biometric Check-in:** Before reaching the airport, passengers enter their biometric characteristics and travel details in the Digi Yatra app. It cross-checks their face data with their travel documents.
2. **Seamless Access:** Passengers arrive at the airport and scan their faces at the entry gates, gaining access if their data conforms with their flight information.
3. **Biometric Authentication for a Touchless Experience:** The integration of security and boarding systems eliminates the need for repeated identity and document checks using facial recognition at security checkpoints and boarding gates (DigiYatra: Future of Air Travel, A Contactless Experience n.d.)
4. **Data Protection:** Every passenger information is stored safely as encrypted data, deleted when travel is done, ensuring confidentiality and compliance with data protection legislation.
5. **Nationwide Scale:** Digi Yatra is well on its way for implementation in most primary Indian airports such as Delhi, Bengaluru and Hyderabad.

The project showcases how AI increases efficiency, sharpens security, and adds convenience for travelers in the air travel segment. It is part of the larger trend toward smart airports enabled by AI and the Internet of Things (IoT).

1. Using the above case, answer the below questions:
2. With respect to global move towards digital transformation, what additional partnerships could the Digi Yatra enter into and create further value for travelers and ecosystem partners?

How does Digi Yatra further facilitate travelers while working directly on data privacy and security issues? What improvements will mitigate privacy concerns in an unplanned outburst?

Source: Ministry of Civil Aviation 2021. https://www.civilaviation.gov.in/sites/default/files/2023-07/Digi%20Yatra%20Policy%20%28DIGI%20YATRA%29.pdf

ETHICAL CONSIDERATIONS

In addition, artificial intelligence can transform the tourism and hospitality sectors. Today's trend of personalization is evolving into hyper-personalization. This raises compelling ethical issues that need immediate attention.

Travelers' personal data is collected and analyzed with facial recognition technology to provide tailored recommendations. Safeguarding people's private rights and establishing expert data-handling processes require stringent regulations to be put in place and goodwill to be nurtured.

There is ambiguity regarding the effect of AI on job positions available in the tourism and hospitality sector. There is a possibility of job loss because a variety of tasks are being automated. The social implications need to be studied, and those displaced need to be provided with reskilling and upskilling solutions to ensure that the transition is inclusive.

In the tourism sector, AI is capable of further propagating existing biases and discrimination. Providing unequal access to holidays, biased recommendations based on race and socioeconomic class can all stem from AI's deep learning. It is critical to eliminate bias that stems from diversity in training data.

The global marketplace is concerned about the culture and ethics of tourism in relation to its transcendental effects on people's lives. Although AI-based virtual tour guides or chatbots can assist with the extreme ease of travel planning, they will definitely not have the spirit and individualistic details that make traveling enriching.

The ethical problems that AI presents within the tourism industry need to be solved in order to espouse an ethical and sustainable approach that benefits the tourists and the locals. In the endeavor to culturalize tourism and sustain credibility, it is crucial to have an equilibrium between technology and human interactivity while maintaining service and cultural relevancy.

In the travel industry, placing trust is already challenging, and if suggestions or actions are carried out in a non-transparent manner, this will result in a reduced amount of trust customers have in the AI system. Users may feel reluctant to utilize the program due to a lack of familiarity with the specific reasoning processes employed by AI systems, which are frequently described as "black boxes."

Reputable AI-based solutions are provided to tourism and hospitality businesses often in regard to best practices that cannot be reasonably expected with little to no infrastructure and funding. This places some regions at a great disadvantage, both financially and socially, cutting many regions off from tourism.

METHODOLOGY

The aim of this study is to determine the impact of AI-driven customer service in the hospitality and travel industry (Sharma et al. 2023). This study involves use of the both primary and secondary data. In addition, a thorough analysis of existing literature was also done. The primary data was collected through a structured questionnaire designed to assess customers' expectations, levels of satisfaction, and perceptions regarding their interactions with artificial intelligence. The secondary data highlights the latest developments in the technological environment within the industry. This research study mainly focused on individuals having familiarity with AI services in the travel and hospitality industry.

The participants were reached through convenience sampling, which netted a total of 130 individuals for the survey. Out of this sample, 118 participants submitted their demographic information, culminating in a complete response rate of 90.76%. The participants of the survey were recruited online through social media and targeted email marketing. As a note, all respondents were informed of the purpose of the study so that their anonymity and privacy could be maintained. The secondary data was collected from scholarly research papers, articles, newspapers, journals, and governmental documentation. These materials substantiated the primary data findings, creating an all-encompassing understanding regarding the adoption and implementation of AI in the hospitality and tourism sector.

RESULTS AND DISCUSSION

This study aims to recognize the customer perspectives on using artificial intelligence. Therefore, a mixed-method approach was used for a thorough comprehension of how users view AI-powered customer service, identifying areas needing enhancement and offering practical recommendations for the industry. The structured questionnaire aimed to capture insights across broad categories, which included the demographic profile of the respondents, the level of similarity and frequency of AI interactions of the respondents, the level of customer satisfaction, the user experience and expectations, and the understanding of ethical considerations and future use.

Demographic Profile

The demographic data was measured by age, gender, and occupation with the help of the computer software SPSS. Results can be seen in Table 1.3. The analysis was completed for the 118 respondents, and the results obtained illustrate the profile of the sample.

Respondents are exclusively younger because the mean age variable value is 1.6186, where 1 represents the youngest age group in the coding system. The coding system uses values ranging from 1.00 to 5.00 to represent five age groups, which include 18–25 years, 26–35 years, 36–45 years, 46–60 years, and 60 years and above. The age distribution demonstrates moderate variability as indicated by the standard deviation of 0.88597 alongside a variance of 0.785. The skewness metric of 1.508 demonstrates a positive skew, which indicates younger age groups have more respondents. The kurtosis score of 1.920 shows a leptokurtic distribution, pointing to respondent clustering within particular age groups.

For the variable gender, the mean value of 1.3983 indicates a higher number of respondents in the first category, i.e. males. The range (1.00 to 3.00) represents three gender categories (male, female, and prefer not to say). The standard deviation (0.54128) and variance (0.293) indicate low variability in gender distribution. The skewness (0.909) indicates a minor positive skew, i.e. the majority is of the first group (male in this case), and the kurtosis (–0.251) is near to a normal distribution, which means a balanced gender spread.

The mean value of 1.8051 for the variable occupation indicates that a greater number of respondents fall into the first occupation category (student). The minimum and maximum values (1.00 and 5.00) correspond to five distinct occupational

Table 1.3 SPSS output for demographics

Descriptive Statistics

	N	Minimum	Maximum	Mean	Std. Deviation	Variance	Skewness		Kurtosis	
	Statistic	Statistic	Statistic	Statistic	Statistic	Statistic	Statistic	Std. Error	Statistic	Std. Error
Age	118	1.00	5.00	1.6186	0.88597	.785	1.508	0.223	1.920	0.442
Gender	118	1.00	3.00	1.3983	0.54128	.293	0.909	0.223	−0.251	0.442
Occupation	118	1.00	5.00	1.8051	1.02331	1.047	1.035	0.223	0.026	0.442
Valid N (listwise)	118									

Source: Author

categories (students, private-sector employees, government public-sector employees, self-employed individuals, and retirees), with the standard deviation (1.02331) and variance (1.047) suggesting greater variability in occupation distribution. The skewness value (1.035) indicates a positive skew, implying that respondents tend to congregate in lower occupational categories, whereas the kurtosis value (0.026) indicates an almost normal distribution, suggesting a well-balanced spread across occupational groups.

In particular, this demographic analysis suggests that the core of the sample appears to be younger, as there is an even gender distribution and varied employment. The skewness and kurtosis values connected with the variables provide further information about the distribution trends of demographic parameters.

AI Interactions

The survey results give insightful information about how well participants understand and engage in AI-based conversations in the tourism and hospitality industry. Graph 1 shows that respondents possess a moderate awareness of AI technologies such as chatbots and virtual assistants used in customer service; a notable portion of participants (31.4%) rated their familiarity at a 3 (where 1 represents the least familiarity and 5 denotes the most familiarity), indicating a moderate level of understanding. These results suggest an increasing recognition of AI technologies, emphasizing their growing significance as tools for enhancing customer experiences.

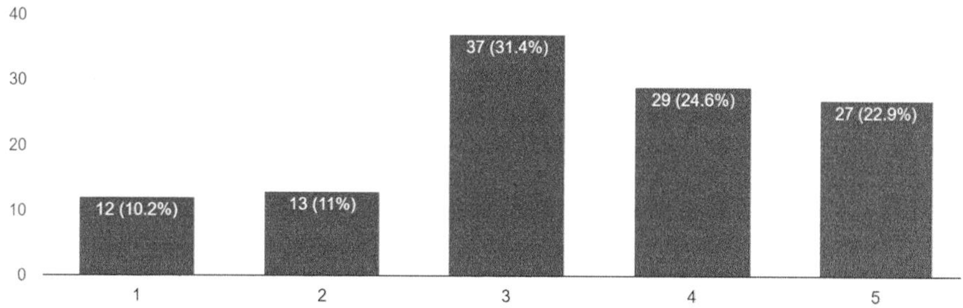

Graph 1 Familiarity of the respondents with AI technologies (e.g. chatbots and virtual assistants) in customer service. (*Source:* Author's Findings).

This casts a positive image of AI adoption, inferring that consumers and industry practitioners are prepared to adopt these innovations. Moreover, there is a call for ongoing education and awareness campaigns in order to help bridge any knowledge gaps and ultimately encourage AI-enabled solutions.

From Graph 2, it can be depicted that the majority of respondents (71.2%) involved in the tourism and hospitality industry have interacted with an AI-powered chatbot or virtual assistant. This shows an increasing acceptance of the industry toward AI-powered customer support solutions. With the high percentage of people who have used AI, it would appear that the industry is becoming more accepting of these technologies. The high usage rate thus implies favorable views of AI in customer support. Firms ought to keep putting money into AI to enhance customer satisfaction and increase operational effectiveness. The design of AI systems must provide individualized help and recommendations based on the unique needs and preferences of individual customers.

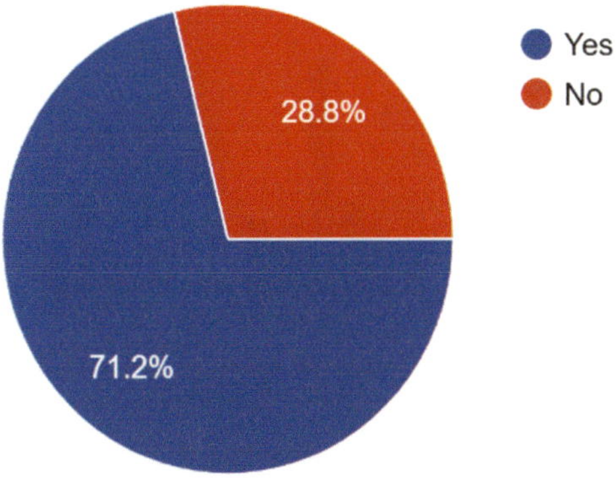

Graph 2 Interaction of the respondents with an AI-powered chatbot or virtual assistant in the tourism and hospitality industry (e.g. hotel booking, travel assistance) (*Source*: Author's Findings).

Customer Satisfaction

As illustrated in Graph 3, respondents were asked to rate the following points to gauge the level of customer satisfaction with AI-powered customer service

***Between overall satisfaction with AI-powered customer service in the tourism and hospitality industry*:** Results from the survey show high levels of overall satisfaction with AI-powered customer service in the tourism and hospitality industry. A very significant majority of respondents (57.62%) were either very satisfied or strongly satisfied with the service. This positive sentiment shows that AI-powered customer service is well received and very capable of meeting the needs of its customers in this competitive setting.

***Broadly, AI-powered customer service has satisfied its function regarding the accuracy of information supply*:** The survey results point to a significant majority among respondents (55.08%) who expressed very high levels of satisfaction. This suggests that AI systems are reasonably accurate in providing information to customers. A smaller proportion, 12.7% of the respondents, reported feeling less satisfied. However, the findings suggest that inaccuracies are not a major problem.

***In terms of personalization, how satisfied are you with the ability of AI-powered services to tailor their responses and recommendations to your needs*:** The distribution of personalization capabilities yielded comparable results, with the majority of participants expressing satisfaction; 58.47% reported being "Very satisfied" or "Extremely satisfied."

***AI-powered services meet your expectations in terms of speed and efficiency*:** Speed and efficiency received the highest proportion of positive responses, with respondents (31%) being "Extremely satisfied" and (34%) being "Very satisfied." A smaller group of respondents (21%) expressed moderate satisfaction, indicating that speed and efficiency are key strengths of AI-powered services.

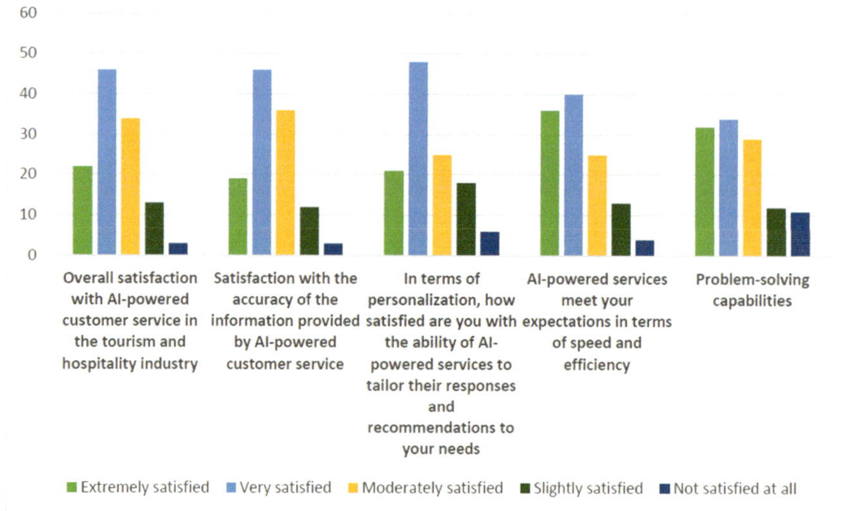

Graph 3 Level of customer satisfaction with AI-powered customer service (*Source:* Author's Findings).

Problem-solving capabilities: As per the survey, there has been a mixed reaction from the respondents. A majority share of respondents expressed their high levels of satisfaction at (62.7%), implying that AI systems perform well in their problem-solving capabilities and are able to satisfy customer problems. However,

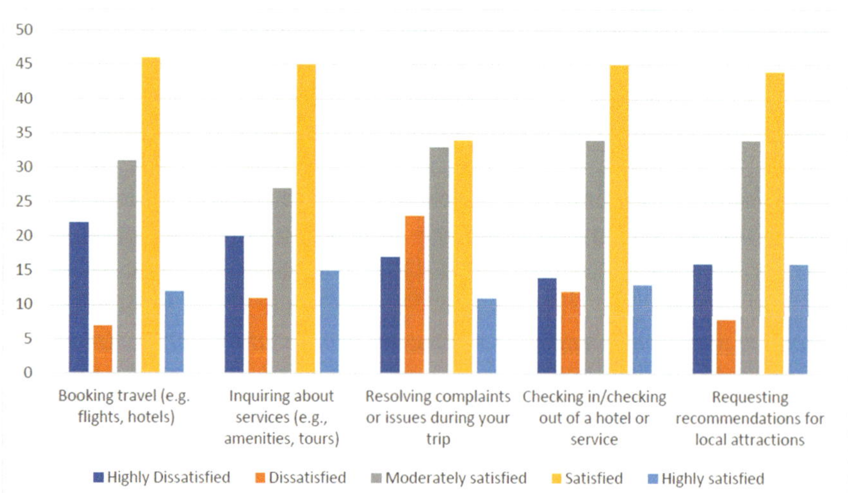

Graph 4 Level of customer satisfaction with AI-powered customer service in different scenarios (*Source:* Author's Findings).

some respondents reported a moderate level of satisfaction (24.57%), while some expressed low satisfaction (19.49%). Therefore, we can say there is still scope for improvement in advancing AI systems to enhance the overall customer experience in resolving individual problem cases.

Further, different scenarios related to travel and hospitality were considered to understand the level of customer satisfaction with AI-powered customer service (Graph 4) in situations such as booking travel (e.g. flights, hotels), wherein the majority of respondents expressed being satisfied (49.1%), while a significant proportion indicated being moderately satisfied (26.27%). Respondents were given another scenario on inquiring about services (e.g. amenities, tours) for which a majority of respondents expressed satisfaction (50.8%) with the AI-powered service's ability to provide information on amenities and tours, while a notable portion indicated moderate (22.8%) and high levels of dissatisfaction (26.27%).

AI-powered services seem to have the greatest difficulty with complaints and issues resolution. A notable portion of respondents exhibited a high level of dissatisfaction (33.89%), which clearly indicates that there has to be an improvement in terms of customer complaint management and solution provision. This degree of dissatisfaction was relatively high. Improvement is clearly required in handling complaints and delivering effective resolutions through AI.

The overall satisfaction levels for check-in and check-out are relatively higher (49.15%) than the number of respondents who expressed dissatisfaction (22.03%). This dissatisfaction indicates that there could be issues with the efficiency of the process faced by customers. Offering more personalization, accuracy, and user-friendliness for the AI tools can enhance customer satisfaction to a greater extent and thus handle this dissatisfaction issue more effectively.

Further, the respondents were inquired about their likelihood of using AI-powered customer service in the tourism and hospitality industry again in the future based on their previous experience (Graph 5), wherein rating 1 means very unlikely and 5 means highly likely.

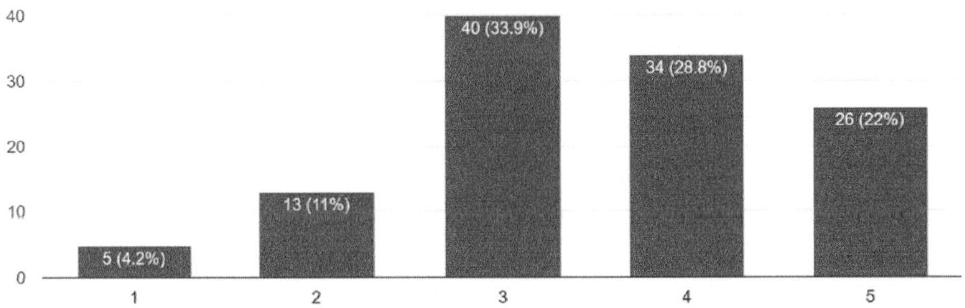

Graph 5 Likelihood to use AI-powered customer service in the tourism and hospitality industry again in the future based on your previous experiences (*Source:* Author's Findings).

More than half of the respondents are positive about attributing AI customer support with reliable attributes. The results show that AI customer care service in the tourism and hospitality sectors is viewed positively by a majority of respondents. A large part of respondents (50.8%), with a significant number, were willing to use these services once more. While fewer expressed ambivalence (33.9%) or a low propensity to use the services again (15.2%), the data leaves no doubt that AI customer care service is responsive to the needs of most respondents.

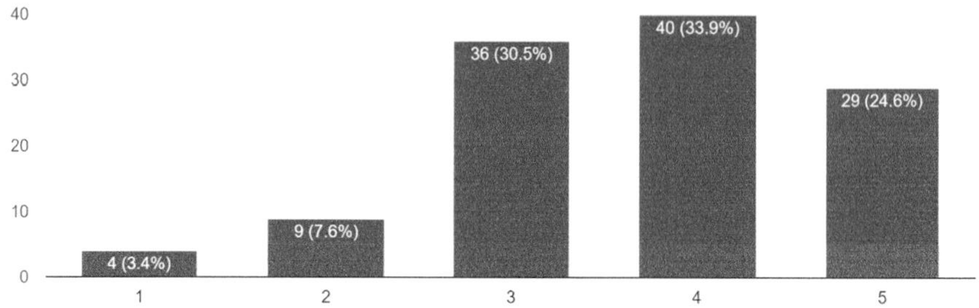

Graph 6 Response time of AI-powered customer service in handling queries (*Source:* Author's Findings).

Maximum of the respondents rated the AI-powered customer service response times reasonably well (Graph 6). A high percentage of 88.98% rated response time at a score of either "good" (3) or "excellent" (4 or 5). The image is still positive, as only a handful of respondents, that is, 11.01%, rated response times to be "fair" (2) or "poor" (1).

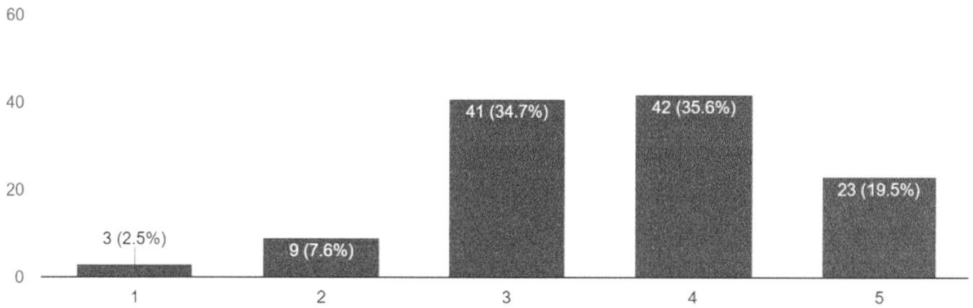

Graph 7 Ease of use of AI-powered customer service in handling queries (e.g. ease of navigation, simplicity of interaction) (*Source:* Author's Findings).

The majority of respondents perceived AI-powered systems as providing fast replies, which would help in customer satisfaction. However, response time increases, particularly for complicated inquiries or during peak times. Promptness is vital, but it should not compromise the accuracy or quality of the response. With these points, the industry can maximize its benefits from using AI.

Graph 7 shows that the majority of respondents found AI-powered customer service, such as ease of navigation and simplicity of interaction, to be relatively easy to use. A significant portion (64.4%) rated the ease of use as "good" (3) or "excellent" (4 or 5). While the overall perception is positive, 10.1% of respondents rated the ease of use as "fair" (2) or "poor" (1). Therefore, by prioritizing user-focused design and ongoing enhancement, the tourism and hospitality industry can fully leverage AI-driven customer service to provide smooth and positive experiences.

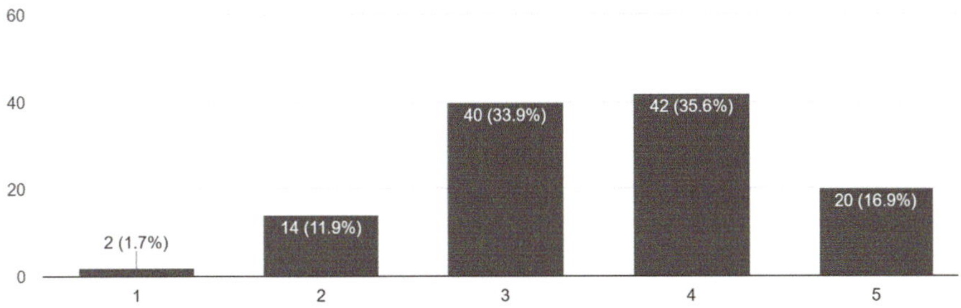

Graph 8 AI-powered customer service enhanced overall travel experience in the tourism and hospitality industry (*Source:* Author's Findings).

The majority of participants believe that AI-powered customer service has had a beneficial effect on their travel experiences (Graph 8). A significant portion (86.4%) rated the impact as "good", "very good", or "excellent". While the overall perspective is positive, 13.59% of participants rated the impact as "fair" or "poor".

A substantial number of respondents (33.9%) find AI-powered customer service to be "much more effective" than human service (Graph 9). Some respondents (6.8%) consider AI-powered and human customer service to be "about the same." Additionally, a few respondents consider it "slightly more effective" (11%). While a significant portion of respondents perceive AI-powered customer service as effective, there is a notable segment that believes traditional human customer service remains superior; i.e. 27.1% and 21.2% of respondents find AI-powered service to be "slightly less effective" or "much less effective," respectively.

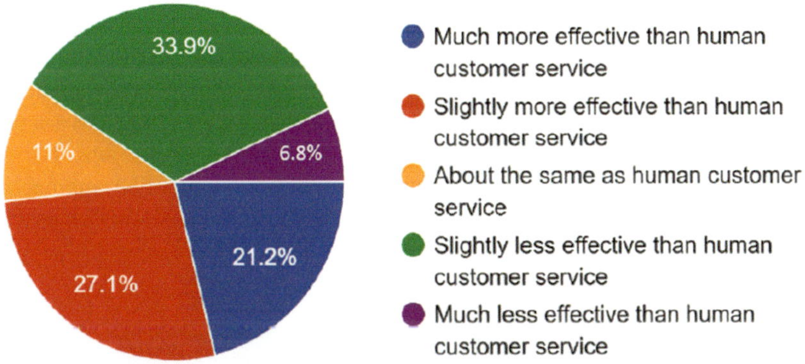

Graph 9 Effectiveness of AI-powered customer service in resolving issues or answering questions compared to traditional human customer service (*Source:* Author's Findings).

In the areas of AI-powered customer services, respondents were most satisfied with the speed of response and 24/7 availability of AI-powered customer service (Graph 10). A significant majority (61%) of respondents were satisfied with the quick response times provided by AI-powered systems. Over two-thirds (68.6%) of respondents appreciated the round-the-clock availability of AI-powered customer service. A substantial number (35.6%) of respondents were satisfied with the professional demeanor of AI-powered interactions. Similar to professionalism, 35.6% of respondents found AI-powered interfaces easy to navigate. A significant portion (33.1%) were satisfied with the accuracy of information provided by AI-powered systems. The least satisfying aspect was the friendliness of interaction, with only 27.1% of respondents expressing satisfaction.

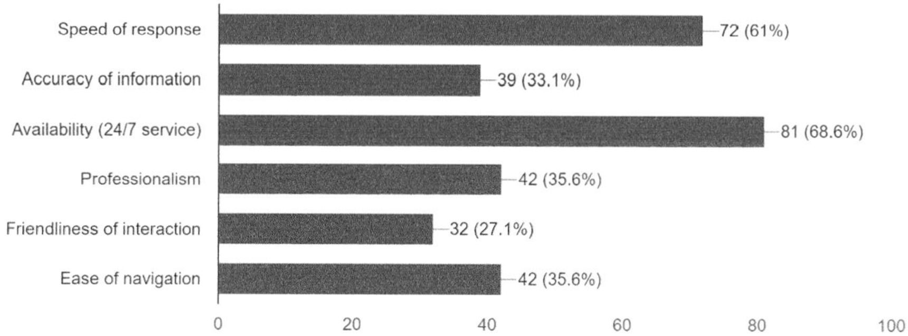

Graph 10 Areas where AI-powered customer service meets or exceeds your expectations (*Source:* Author's Findings).

User Experience and Expectations

The survey highlights several perceived aspects of AI-powered customer service by the respondents in the tourism and hospitality industry that need significant improvement (Graph 11), including personalizing responses to individual needs (54.2%), handling special requests or unique situations (52.5%), and understanding complex queries (46.6%), followed by maintaining a human-like tone or empathy (33.1%) and ensuring data privacy and security (29.7%). It can be inferred that customers are looking forward to more personalized solutions and interactions from AI systems. These systems should be capable of handling a variety of customer cases and adapting from these different customer interactions. For this, AI systems should handle customer queries in a more human-like approach by improving the accuracy and efficiency of natural language processing algorithms.

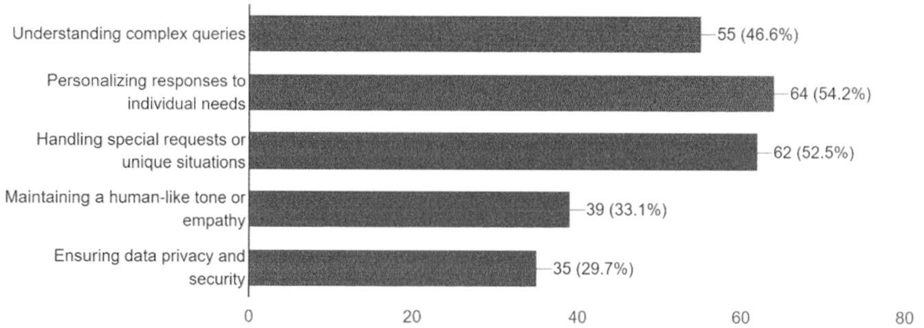

Graph 11 Aspects of AI-powered customer service in the tourism and hospitality industry that needs significant improvement (*Source:* Author's Findings).

To understand which type of AI features encourage the usage of AI-powered services (Graph 12) more amongst the respondents, the survey highlights that respondents are primarily motivated to use AI-powered services by two key factors. First, on the basis of faster response times (46.6%), indicating that rapidity and effectiveness are vital for the implementation of AI in the tourism and hospitality sector. Customers appreciate prompt and timely support, which AI systems can offer through automated systems and quick response rates.

Secondly, improved accuracy of responses (22%) emphasizes the significance of providing accurate information through AI systems. Other key factors highlighted by the respondents include the integration of AI-powered services with human agents (16.9%) and a focus on more personalized services (14.4%). This suggests establishing the right equilibrium between the efficiency and promptness of these AI-powered services so that customers experience robust AI that provides accurate and regularly updated data, along with eliminating the possibility of inaccuracies and ambiguity.

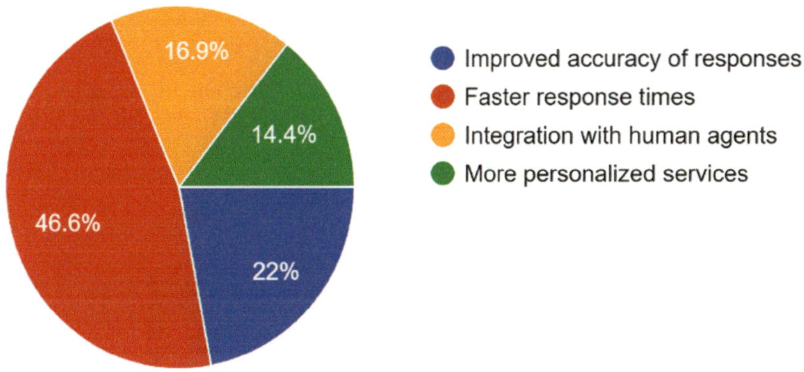

Graph 12 AI features encouraging the use of AI-powered services more often (*Source:* Author's Findings).

The respondents' answers were analyzed to understand the most significant drawbacks associated with AI-powered customer service (Graph 13), which include difficulty in handling complex issues (48.3%) as a major concern and lack of empathy and human touch (44.1%), emphasizing the importance of maintaining a human touchpoint in customer services while using AI. It can be inferred that the complexity of customer issues poses a challenge for companies, as current AI systems are not sufficiently trained to manage intricate and nuanced inquiries. Therefore, ongoing advancements in AI's ability to comprehend and address customers in the most appropriate manner are necessary. The significance of human touchpoints during the entire customer service process cannot be underestimated, as customers genuinely appreciate a personal touch during their interactions, and hence, it is not feasible to completely eliminate human profiles with AI systems.

Also, privacy and data security concerns (42.4%) and over-reliance on technology (30.5%) were the other major drawbacks according to the respondents.

Trust and transparency are critical components to mitigate concerns regarding privacy and data security. It is, therefore, important to establish trust for the effective integration of AI into customer service.

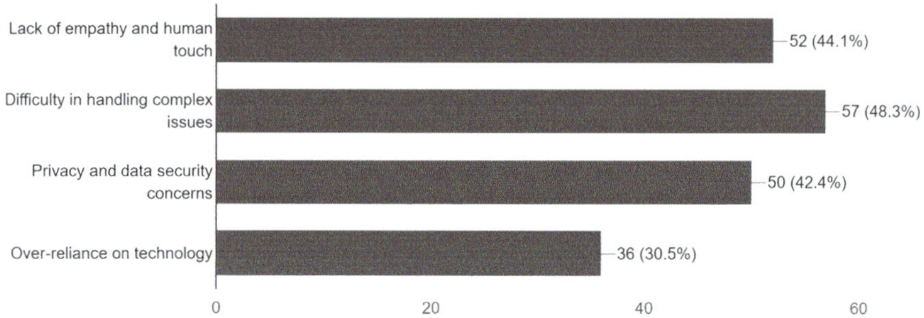

Graph 13 Most significant drawback of AI-powered customer service (*Source:* Author's Findings).

Ethical Considerations and Future Use

The results indicate how people view the future function of AI in customer support, emphasizing its potential and effects in Graph 14.

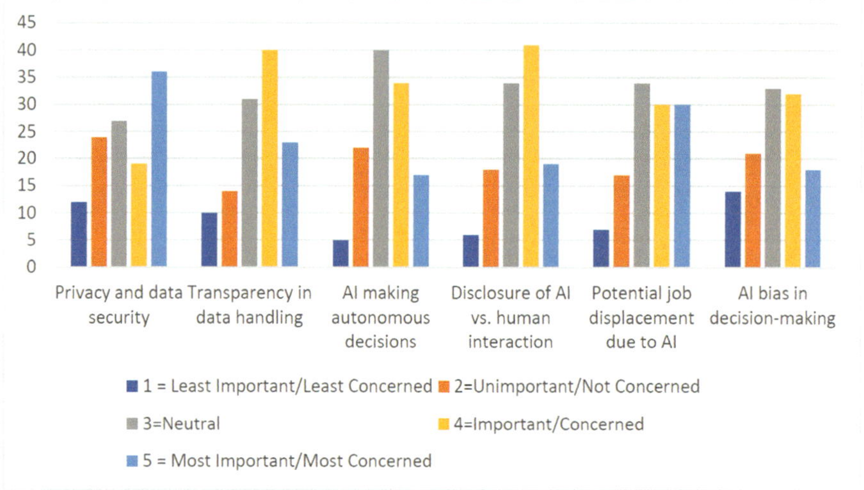

Graph 14 The importance or concern for the different aspects of AI-powered customer service (*Source:* Author's Findings).

Privacy and Data Security

The concerns of privacy and data security have emerged as significant issues in the survey, with a considerable percentage of respondents deeming them "important" (16.1%) or "most important" (30.5%).

Transparency in Data Handling

Transparency is recognized as another crucial challenge, as many respondents (53.3%) placed high significance on it, which implies there should be detailed policies regarding the collection, use, and dissemination of data. Customers anticipate that organizations will be open about their practices and hold accountability.

AI Making Autonomous Decisions

Despite evident concerns surrounding AI's capacity for autonomous decision-making, a notable portion of the population remained neutral in their views (34%). Only a small number expressed strong apprehensions. This illustrates a split opinion: some individuals have faith in AI's decision-making abilities, whereas others harbor doubts about ethical implications and fairness.

Disclosure of AI vs. Human Interaction

The respondents had ethical concerns related to the clear disclosure of whether they are interacting with a human or an AI system. The majority of participants identified this as a relevant concern (50.8%). Participants look forward to more transparency and demand to be informed prior to any communication.

Potential Job Displacement Due to AI

The major concern that has been anticipated by the participants is the potential replacement of jobs due to AI (50.8%). This is another ethical issue.

AI Bias in Decision-Making

The participants reflected their concerns on the perpetual and probable bias in the decision-making process by AI systems, perhaps unknowingly, as another ethical challenge. The respondents highlighted the significance of non-discriminatory decision-making (42.3%) within AI systems.

Graph 15 shows the findings pertaining to the likelihood or impact of the issues and priorities concerning various ethical aspects of AI-enabled customer service.

***In the next 5 years, customer support will be handled predominantly by AI*:** More than half of the respondents (53.8%) expect AI to be the supreme force in customer service, with the majority showing optimism about this shift. Furthermore, quite a few participants held neutral views (27.11%), while others voiced skepticism regarding the potential for AI to assume such a dominant role (19.4%). These diverse reactions suggest that while there is broad acceptance of AI capabilities, hesitance remains concerning its quick integration into customer service contexts.

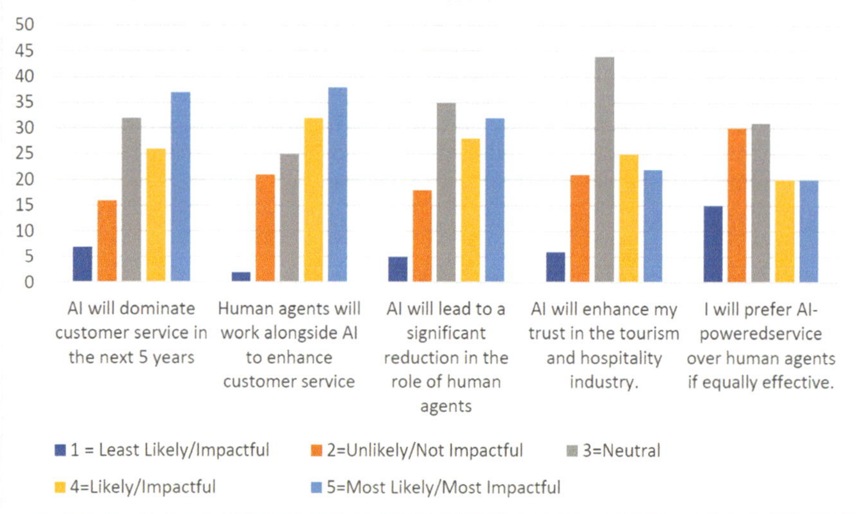

Graph 15 The likelihood and impact of the different aspects related to the future use of AI in customer service (*Source:* Author's Findings).

AI will empower human agents to deliver a superior customer service experience: The strongest participants' consensus was recorded here, with the majority agreeing that AI will combine the effective skills of customer service representatives and technology (59.3%). This shows that they preferred the blended approach in which humans and AI come together to improve service as opposed to one replacing the other.

Human agents will be replaced in great numbers due to AI: There is a strong belief that AI can reduce the reliance on human agents; however, many respondents also claimed neutrality (30%). It can be inferred that though some are moving towards automation, others are reluctant to adopt AI, as the human touch cannot be replaced.

***AI will enhance my trust in the tourism and hospitality industry*:** The overall reaction seems to be mixed on whether AI systems would lead to increased trust or not. About 37.2% of the respondents expressed uncertainty. Building trust in AI systems remains a significant challenge since transparency, accountability, and reliability serve as essential ethical considerations.

I will prefer AI-powered service over human agents if equally effective: A minority of survey participants preferred AI solutions to human agents at a rate of 33.8%, despite AI systems delivering equal effectiveness. It can be inferred that customers are not fully comfortable with AI-powered services and require a human touch instead.

The ethical framework governing the use of AI systems in customer service hinges on principles like privacy protection alongside transparent and equitable practices. The operational efficiency improvements brought by AI require stakeholders to address data protection challenges as well as bias and job loss issues to build trust and secure equality while supporting responsible societal adoption. The saying "great power brings great responsibility" reminds us that ethical discussions about artificial intelligence implementation in tourism and hospitality sectors must address the human-machine interaction balance.

CONCLUSIONS

Developments in the future of tourism are expected to be influenced by AI technologies. Therefore, further investment in research and development is needed to improve the effectiveness of AI systems. While there is greater automation, people will still be needed where empathy, imagination, creativity, and delicate decision-making are required. The interaction of AI with human support will result in greater efficiency and fulfillment of customer needs. This will not only make services more effective and satisfactory but will also foster frequent feedback from customers that will pinpoint the changes needed for improvement in customer service to ultimately build brand loyalty.

REFERENCES

Apriorit. 2023. Rule-based Chatbot vs AI Chatbot: Which to Choose for Your Business? https://www.apriorit.com/dev-blog/web-ai-vs-rule-based-chatbots

Barney, N. and Mixon, E. 2024. What is customer experience (CX) and why is it important? Search Customer Experience. https://www.techtarget.com/searchcustomerexperience/definition/customer-experience-CX

Barten, M. 2024. How Artificial Intelligence is Changing the Travel Industry. Revfine.com. https://www.revfine.com/artificial-intelligence-travel-industry/

Bradley, T. 2024. How AI is revolutionizing customer experience. Forbes. https://www.forbes.com/sites/tonybradley/2024/10/11/how-ai-is-revolutionizing-customer-experience/

Chen, Y. and Prentice, C. 2024. Integrating artificial intelligence and customer experience. Australasian Marketing Journal (AMJ). https://doi.org/10.1177/14413582241252904

Dang, T.D. and Nguyen, M.T. 2023. Systematic review and research agenda for the tourism and hospitality sector: co-creation of customer value in the digital age. Future Business Journal. 9(1). https://doi.org/10.1186/s43093-023-00274-5

Filipsson, F. 2024. Top 10 best AI Chatbots for Customer Service. Redress Compliance—Just another WordPress site. https://redresscompliance.com/top-10-best-ai-chatbots-for-customer-service/

KLM. 2017. KLM Welcomes BlueBot (BB) to Its Service Family. https://news.klm.com/klm-welcomes-bluebot-bb-to-its-service-family/#

Maya—AI assistant for travel companies. (n.d.). https://www.mayatravel.ai/

Ministry of Civil Aviation. 2021. Digi Yatra Biometric Boarding System DY-BBS. In Ministry of Civil Aviation. https://www.civilaviation.gov.in/sites/default/files/2023-07/Digi%20Yatra%20Policy%20%28DIGI%20YATRA%29.pdf

Omind. (n.d.). Personalization in CX: Crafting unique customer journeys. https://www.omind.ai/blogs/personalization-in-customer-experience-tailoring-the-journey-for-each-customer

Qualtrics. 2023. The top customer service metrics to measure Qualtrics. https://www.qualtrics.com/en-au/experience-management/customer/service-metrics/

Radwan, A. 2024. Chatbots and virtual assistants: AI transforming routine business operations—ProfileTree. ProfileTree Web Design and Digital Marketing. https://profiletree.com/chatbots-and-virtual-assistants/

Shah, P. 2023. Travel Chatbot Case Study—Amtrak's Chatbot, Kevit Technologies. Kevit Technologies. https://kevit.io/travel-chatbot-case-study-amtraks-chatbot/

Sharma, S., Rawal, Y.S., Soni, H. and Batabyal, D. 2023. Technological Impacts of AI on Hospitality and Tourism Industry. *In*: Saraswat, M., Chowdhury, C., Kumar Mandal, C., Gandomi, A.H. (eds). Proceedings of International Conference on Data Science and Applications. Lecture Notes in Networks and Systems. Vol. 551. Springer, Singapore. https://doi.org/10.1007/978-981-19-6631-6_6

Takyar, A. 2023. Exploring AI in customer support across industries. Leeway Hertz—AI Development Company. https://www.leewayhertz.com/ai-in-customer-support/

Wong, S., McClain, E., Cohn, J. and Visser, R. 2021. Airport biometrics: A primer. *In*: Transportation Research Board eBooks. https://doi.org/10.17226/26180

FURTHER READING

Awad, A.I., Babu, A., Barka, E. and Shuaib, K. 2024. AI-powered biometrics for Internet of Things security: A review and future vision. Journal of Information Security and Applications. 82: 103748. https://doi.org/10.1016/j.jisa.2024.103748

Hilton. 2016. Hilton and IBM pilot "Connie," the world's first Watson-enabled hotel concierge. Hospitality Net. https://www.hospitalitynet.org/news/4074810.html

Team, E.S. 2024. AI in tourism and hospitality: The essential guide - online business school. Online Business School. https://esoftskills.com/ai-in-tourism-and-hospitality-the-essential-guide/

USA Today. 2016. Introducing Connie, Hilton's new robot concierge. USA TODAY. https://www.usatoday.com/story/travel/roadwarriorvoices/2016/03/09/introducing-connie-hiltons-new-robot-concierge/81525924/

Chapter **2**

Applying Artificial Intelligence to Reduce Food Waste and Strengthen the Sustainable Economy

Dharmender Singh Rawat*[1] and Rakesh Dani*[2]

[1]Research Scholar, Department of Hospitality Management, Graphic Era Deemed to be University, India

[2]Professor, Department of Hospitality Management, Graphic Era Deemed to be University, India

INTRODUCTION

Over the past two decades, organic waste, particularly food waste, has developed into a critical issue with substantial ecological, social, and economic impacts worldwide. Food waste is a foremost contributor to global greenhouse gas emissions, ranking third after the U.S. and China, with a projected 3.3 gigatons of emissions annually (Mganga et al. 2021). The Food and Agriculture Organization (FAO) reports that approximately 1.3 billion tons of food, or one-third of the food produced for humans, is either lost or unused annually. This staggering revelation has sparked calls for action from grassroots organizations, global authorities, and nonprofit establishments. As food continues to be lost and wasted, food prices rise, exacerbating global food insecurity and poverty (Canton 2021). Waste is one of the most significant challenges confronting the younger generation. As

*For Correspondence: d.s.rawat07@gmail.com

DOI: 10.1201/9781003540083-2

countries around the globe develop, their responsibility for maintaining a safe and healthy environment increases. While developed and developing nations grapple with waste management issues, emerging countries are devising innovative solutions that yield substantial positive outcomes (Thi et al. 2015). The European Commission 2014 has categorized food waste into three main groups: (i) food losses, which occur during the production stage; (ii) unavoidable food waste, such as fruit cores and banana peels, which are lost during the consumption stage; and (iii) food waste that could have been prevented, such as food that was lost during the eating stage. These categories are based on diverse phases of the food chain (Gustavsson et al. 2011). Losses and waste occur at every stage of the food chain, including production, shipping, warehousing, processing, retail, and consumption. As part of the "Agenda 2030 for Sustainable Development," the international community has vowed to address food scarcity and malnutrition by reducing food loss and waste (Delgado et al. 2017). High-income nations throw away an average of 79 kg of food per person per year, according to the United Nations Food Waste Index Report (UNEP Food Waste Index Report 2021).

For poor countries especially, controlling food waste is perhaps a more important problem than for industrialized ones. Food waste becomes a serious environmental problem since inadequate separation from municipal solid waste (MSW) results in increased greenhouse gas (GHG) emissions in landfills. Food waste generates greenhouse gas emissions when it breaks down, thereby aggravating climate change (Isabelle 2014). The Greenhouse Effect is the mechanism by which certain compounds trap heat by absorbing and releasing radiation within the Earth's atmosphere. Among the leading greenhouse gases are nitrous oxide (N_2O), ozone (O_3), carbon dioxide (CO_2), methane (CH_4), fluorinated gases (F-gases), and water vapor (H_2O). Burning coal, oil, and gas, deforestation and land-use alterations, increasing cattle and agriculture, industrial operations and manufacturing, transportation, and energy generation are among human activities that help to produce these gases (Munesue et al. 2015). Factors such as population increase, higher individual consumption, technological developments, mass displacement, and ease of transportation have created environmental issues and resource depletion. These problems manifest in soil degradation, worsening air and water quality, carbon emissions, climate change, and global warming. Food waste has also grown to be a worldwide issue, aggravating already existing problems.

While many people around the world face food scarcity, vast quantities of food are wasted every day in other regions (Frayne et al. 2010). Each year, diseases related to hunger result in the deaths of approximately 10 million people. Globally, around 1.3 billion tons of food are lost or wasted every year, which could feed 870 million people. In developing nations, over 140 million children are underweight (Charney 2009). In the upcoming years, the hotel industry is anticipated to experience significant growth across various regions globally (Berman 2014). As the hospitality industry expands, it is expected to generate more waste. The costs associated with solid waste in the hotel industry are influenced by several factors, including labor expenses, waste disposal, and transportation (Pirani and Arafat 2016).

FOOD WASTE AND THE SUSTAINABLE ECONOMY: CURRENT SITUATION

A big issue is that an estimated one-third of all food produced globally vanishes or becomes worthless. With around USD 936 billion yearly, this waste causes a significant financial loss. Apart from the financial impact, food waste has major social consequences as well as ecological ones, including resource depletion and emissions of greenhouse gases (Jamaludin et al. 2022). Along with improvements in production techniques and eating behavior, decreasing food waste is critical for enabling sustainable development inside the food system while presenting a destination wisely (Batabyal 2010, Jamaludin et al. 2022). We can contribute to minimizing the negative environmental consequences of the food system by lowering food waste and raising food and water safety (Ndondo 2023). Long-term effects of food waste for the economy, society, and the environment abound. Aiming to reduce waste, a sustainable economy stresses the effective and optimal utilization of resources. Maximizing the value of products, minimizing resource consumption, and conserving the long-term worth of materials and goods can help to continuously reduce waste (Tamasiga et al. 2022). As per the Food and Agriculture Organization (FAO), around one-third of all food generated for human consumption is thrown away or lost. The FAO claims that local elements influence the reasons for food waste, which differ depending on the nation. After harvest, food loss in low-income nations usually occurs in the manufacturing, handling, storage, and processing phases. Budgetary restrictions, outdated equipment, and inadequate infrastructure account for most of this waste. Administrative and

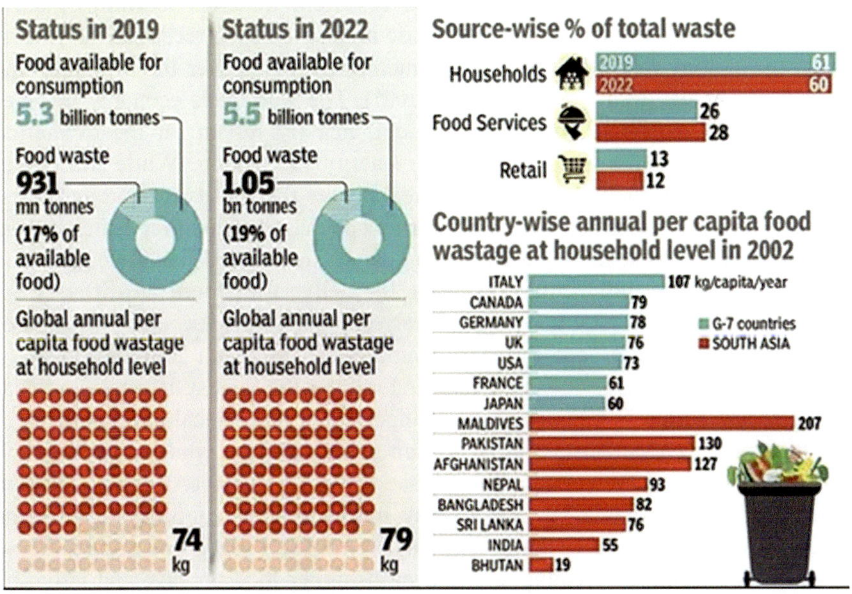

Figure 2.1 Global food waste statistics between 2019 and 2022 (*Source:* UNEP Food Waste Index Report 2024).

technical difficulties also play a chief role in the total amount of food lost (Thi et al. 2015). Figure 2.1 shows the global food waste statistics of Household,Food Service and Re The dissemination and consumption stages of the food value chain are where most of the food waste occurs in developed and developing countries. The amount of food wasted is also impacted by regional consumer behavior and government regulations in both wealthy and underdeveloped nations (Van der Werf and Gilliland 2017, Batabyal and Das 2022). Many governments, corporations, and individuals are actively working on lessening waste and encouraging more enduring food systems, exhibiting a rising consciousness and dedication to addressing the problem of food waste. Research increasingly supports implementing a sustainable economy to effectively tackle food waste (Kumar et al. 2022). In the end, implementing food waste reduction and building a sustainable economy will require the ongoing cooperation and dedication of all parties in the food chain, as well as constant innovation and adaptability to changing situations and new challenges (Calvo-Porral et al. 2017). The following figure shows annual household food waste statistics from a few chosen nations, illustrating the extent of food waste worldwide.

AI'S POTENTIAL TO REDUCE FOOD WASTE AND PROMOTE A SUSTAINABLE ECONOMY

The application of computers to emulate human cognition is termed artificial intelligence (AI). AI represents a dynamic template that significantly influences the world of science, governments, and conventional governmental and financial methods addressing the concern. Broad research, machine learning, picture recognition, and interpreting common prose are some of the recent developments in AI technology that imply these instruments will carry over the influence and change day-to-day living (Sharma et al. 2021). The worldwide economy operates on a "take, make, and dispose of" model, heavily reliant on the extraction of limited resources and fossil fuels for energy production. While this linear approach has generated unprecedented wealth over the past two thousand years, it has also caused significant harm to our biosphere. AI, which replicates human cognitive functions such as the ability to analyze and review, could change food systems and global production, including the delivery of food. AI differs from human intelligence in that it does not imply computers possess consciousness or intelligence. Instead, it denotes that a computer can resolve a certain problem or group of similar ones. The application of AI to alter the food system from a linear to a circular design is one of the most important technological developments of the present time (McKinsey 2019). AI can generate value while protecting and restoring biological systems. Research done by the Ellen MacArthur Foundation discovered that AI has a big role to play in changing our food system for the better! It can help in three important ways: first, by cutting down on food we don't need; second, by creating and sharing healthier food options; and lastly, by helping us find local and sustainable foods whenever we can (Ellen Mac Arthur 2015). As per the research of McKinsey Global Institute, by the year 2030,

AI might boost the global economy by about USD 13 trillion! It is highly impactful. Plus, if we use strategies to cut down on food waste, AI could even help businesses make up to USD 127 billion more each year, which represents a substantial financial gain (McKinsey 2019). With the rise in demand for quick and easy meals, food processing is changing all the time. People who think outside the box, like innovators and designers, are using AI. They want to help us find ingredients from plants that are farmed in a good way. Also, they're working on switching out animal proteins for tasty plant-based options. Plus, they're focusing on reducing waste during processing and getting rid of harmful additives. This whole idea is about making healthy food available and sustainable for everyone (Ellen Mac Arthur 2015). AI has a huge prospect of cutting down on food waste and boosting sustainability in food production. For the food industry, it's a big opportunity to support the sustainable economy as well. By using AI technology, businesses can make better decisions, simplify processes, and create new solutions to meet urgent challenges facing our universal food chain.

FOOD WASTE REDUCTION WITH AI IN HOTELS GROUP

In the next few years, the hotel industry is expected to experience significant growth across different regions of the world (Berman 2014). In fact, tourism played a vital role, contributing 9.1 percent to global GDP in 2023. This means that the overall impact of travel and tourism on the global economy was almost 10 trillion U.S. dollars that year. In 2020, there was a notable decline—travel dropped by 75%. But it is expected, by the end of 2024, that things will bounce back completely. Looking ahead, we see domestic traveler nights rising at about 3% each year, which should total around 19 billion by 2030. Foreign traveling is also predicted to return to its historical average of nine billion nights within the same timeframe. Travel expenditures are forecasted to follow a similar upward trend, with an expected total of $8.6 trillion in 2024, accounting for approximately 9% of the global GDP for that year. People continue to enjoy traveling, seeking fresh experiences in diverse locations. But where do travelers typically start their journeys, and where do they go? To provide insights into current travel patterns and growth projections through 2030, we have analyzed travel flows. For this report, we have identified four key regions: the Americas, Africa, Europe, Asia, and the Middle East. India's rapid economic expansion, with a 6% annual GDP growth, is fostering a new generation of travelers. This growth is expected to lead to a 9% annual increase in travel spending by 2030. To meet the anticipated demand, Indian lodging and airline companies are making substantial investments. In the foreseeable future, The Indian Hotels Company Limited plans to open two new hotels every month by increasing the pace of its hotel pipeline. In 2024, seven hotel chains will introduce new brands in India, demonstrating that global players are not staying put. These include new properties under the Hilton Curio and Tapestry brands, as well as Marriott's first Moxy and Tribute brands in India. Aside from these numerous small and large hotel chains, more are on the way.

The emphasis on development has shifted from large metropolises like Delhi and Mumbai to smaller, rapidly developing cities like Chandigarh, Hyderabad, and other small cities (Ellen Mac Arthur 2015).

My recent visit to the Nainital district in the Bhowali region, a small hill station known for its lush greenery, pleasant weather, and fruit cultivation, shows how the tourism sector is growing day by day. It's always peak time during summer, long weekends, and festival seasons. The Nainital district is famous as a tourist destination. Local Government authorities of Uttarakhand (Urban Municipal Solid Wastage Management) have taken commendable steps to dispose of hotel and resort waste, which was earlier thrown and dumped indiscriminately by different hotels, resorts, and restaurant chains. A new garbage disposal machine was installed, and a proper area was developed where the garbage is dumped and further processed through this AI-based garbage disposal plant.

Figure 2.2 Hillside utility shed structure
(*Source:* Photographed by Researcher).

Figure 2.2 shows the garbage disposal machine installed in Nainital district of Bhowali region by Local government authorities of Uttarakhand. As tourism grows, more hotels will be occupied, and due to this, food waste will also grow. Hotels and restaurants inherently generate substantial food waste due to their operational nature. This issue is a growing concern for the government and relevant authorities. To ensure that these establishments operate efficiently, implementing control measures is essential. Food waste in hotels and restaurants happens at many stages. This is frequently due to demand fluctuations, overstocking, and unproductive techniques. Poor communication among employees, behavioral issues, and inexperience in food preparation also contribute. Over-merchandising and expired products play a role, too (Pirani and Arafat 2016).

Once food gets eaten, waste can still pile up. This can come from large portion sizes, inefficient service techniques, and consumer likes/dislikes. The kind of waste in hotels and restaurants depends on the ingredients being used.

Common items that end up as waste include packaging materials, leftover food, bones, eggshells, and potato peelings (Filimonau and Delysia 2019). Managing food waste is a complex dilemma because of the many factors at play. Data shows several harmful effects on the environment, economy, and societal well-being. Unfortunately, there hasn't been an ideal solution for solid waste management in hotels and restaurants yet. Still, stakeholders are actively working to tackle the issues that come with food waste in these places (Amicarelli et al. 2022). Many hotels are now using AI technology to cut down on kitchen waste. For instance, the Intercontinental Hotels Group (IHG), which is the largest hotel chain globally, has teamed up with Winnow Solutions. This partnership aims to integrate artificial intelligence into hotel management systems, helping to better manage food waste efficiently.

Figure 2.3 Waste management and recycling facility (*Source:* Photographed by Researcher).

This proposal aims to reduce food waste by as much as 30%. Figure 2.3 show the garbage is kept near the installed garbage disposal machine in Nainital district of Bhowali region for the further process. AI technology is beneficial because it streamlines operations by reducing the need for extensive staff training, automates data entry processes, and enhances transaction accuracy compared to manual methods. By connecting a kitchen trash can to a tablet, food waste data can be collected from the kitchen. This technique lets the kitchen crew quickly find and record what is thrown out. Winnow's system detects foods automatically without human involvement. After that, the acquired information is used to customize menus, enhance buying decisions, and change cooking methods for forthcoming events. Currently, seven IHG hotels spread around Europe, the Middle East, Africa, and Asia have implemented this technology, with plans to expand to thirty more sites in the near future.

IHG is also investigating bringing this technology to the Greater China area and North America. Chief Executive Officer for EMEAA Kenneth Macpherson of IHG underlined the company's dedication to ethical practices. With more than 5,600 hotels worldwide, IHG has a major global presence, he noted. He praised the continuous attempts to have a greater impact by cutting plastic straws and implementing creative solutions to reduce environmental effects. Aimed at producing enticing, low-waste menus for different dining situations, Macpherson expressed enthusiasm about the cooperation with Winnow. He observed that a handful of hotels in the Middle East are pioneering this initiative and have already seen substantial reductions in food waste. The CEO and creator of Winnow, Marc Zornes, emphasized that misuse of food is a worldwide issue that impacts kitchens everywhere. Kitchens frequently throw out more food than they realize when they don't have an adequate system in place to track waste. He was proud to have worked with the cutting-edge hospitality company IHG to assist them in becoming the first significant global hotel brand to use artificial intelligence to improve their sustainable food and beverage operations and drastically reduce food waste (Thakar 2019).

UTILIZING AI FOR RECYCLING AND WASTE MANAGEMENT

Effective waste management is essential for reducing or eliminating trash and maintaining hygiene. While proper garbage collection and disposal are crucial components, good waste management encompasses more than just recycling and waste disposal. One of the key benefits of recycling is that it decreases our reliance on raw materials, contributing to both efficient waste disposal and resource conservation. When regenerative methods aren't possible, artificial intelligence (AI) has a fundamental task in finding the most cost-effective and suitable dumping methods for different returning goods (Ihsanulla et al. 2022).

Figure 2.4 Samur AI robot technology
(*Source:* SAMURAI Sorting Robot Machine-X (2025)).

Figure 2.4 shows that how Samur AI robot Machine segregate the recyclables including plastic bottles,boxes and container. Through its AI sensor it pics 70 recyclables products per minute. Recently, new methods—like AI—have gained more focus on resource recovery. This includes recycling, reusing, and even generating energy from waste materials. AI improves decision-making in biowaste management and the production of biomass energy by considering environmental, social, and financial aspects related to waste handling. Ultimately, this strategy helps turn waste into a resource that holds value (Aniza et al. 2023).

Recycling, reusing, or recovering materials before they are no longer essential is where artificial intelligence (AI) technologies are important. It reduces the overall consumption of natural resources in waste control. Reusing items or recycling them will help us greatly reduce the demand for resources. Here is where artificial intelligence excels and helps to promote environmentally friendly garbage management (Andeobu et al. 2022).

One significant development in this area is the Samur AI robot design produced by Machinex. In the recycling industry, this creative solution is unique. It first surfaced at the Waste Expo exhibition in Las Vegas. This device has fascinating possibilities that could transform our approach to better recycling. Designed to recognize several objects on a moving conveyor belt, the Samur-AI robot works especially well for separating recyclables, including plastic bottles, boxes, and containers. The robot employs a suction mechanism once it finds an object to extract it and puts it in a suitable container. With up to 70 picks per minute, the Samur-AI greatly outperforms human teams in terms of performance.

This advancement enhances overall productivity and substantially reduces operating costs. Applying intelligent computerization methods to waste management is both practical and time-efficient. These approaches use advanced machine algorithms to improve adaptability. By leveraging experimental sensor data from light transmission and sound testing, they can distinguish between different types of objects. Additionally, sound waves are used to effectively separate materials like glass, metal, and plastic, while electromagnetic sensors are employed to filter out real metal. Many companies have developed robots that utilize machine learning for these purposes. Firms such as Biotech, Intel, and Zen Robotics are actively investing in and developing technologies for efficient waste disposal (Ahmed and Asadhulla 2020).

A SUSTAINABLE ECONOMY INITIATIVE SUPPORTED BY AI AND ITS POTENTIAL BENEFITS

Resource recovery from waste is getting a lot of attention these days. Extracting the most value from resources is crucial. Artificial intelligence plays a key role in this effort. It helps adopt environment-friendly production techniques. Energy efficiency can be improved through AI, too. Products and components can last longer with their help. AI is great at monitoring and tracking items in real time. This allows for assessing their residual value effectively. As a result, we see a boost in sustainable economic practices. Moreover, this technology supports decision-making at the company level, especially regarding materials and processes. Sustainable AI means building artificial intelligence systems that help protect our environmental resources. This is for both today's people and those in the future;

it also supports financial frameworks that are good for society while respecting community values (Van-Wynsberghe 2021). Artificial intelligence can check and assess items in real time. This helps to figure out their remaining worth, which encourages sustainable economic practices. As a result, manufacturing companies can make better decisions regarding materials and processes. Artificial intelligence employs rapid methods to enhance system adaptability and securely integrate information into operational processes, addressing the challenges of transformation (Mboli et al. 2022). AI can help in the development of visual aids that effectively summarize data flows associated with resources, goods, and processes. These tools can help uncover previously unknown benefits of a sustainable economy (Vinuesa et al. 2020). Table 2.1 highlights how AI applications contribute to reduce food waste management. It demonstrates AI's role in monitoring food quality, optimizing inventory, improving demand forecasting,enabling surplus food redistribution and enhancing consumer awareness.

Table 2.1 AI and food waste: Minimize waste waste and providing food for future generations

S. No.	AI Application	Food Waste Management
1.	Astute observation and analysis	To minimize waste, proactive measures can be implemented by utilizing AI-enabled sensors and data analytics to continuously supervise food quality, expiration dates, and storeroom circumstances in real-time.
2.	Analytical Forecasting	Artificial intelligence algorithms leverage data analysis to predict customer demand, optimize inventory management, reduce overproduction, and minimize waste throughout the supply chain.
3.	Smart Inventory Management	An AI-powered inventory system optimizes stock levels, reduces food spoilage, and improves supply chain efficiency, all the while conserving resources and minimizing waste.
4.	Knowledge and Awareness of Consumers	AI-driven applications and platforms offer personalized advice on meal preparation, proper storage, and creative ways to reduce food waste at home.
5.	Redistribution and donation	AI reduces waste and fights food poverty by identifying excess food and connecting it with organizations that can distribute it to those in need.

Source: (Onyeaka et al. 2023)

AI-POWERED MONITORING AND OPTIMIZATION OF FOOD PRODUCTION, AGRICULTURE, AND SUPPLY CHAINS

Food processing techniques, longer shelf lives, and nutritional content are issues that consumers are continuously worried about. Due to AI's multidisciplinary character, it may be employed to supervise flows and do any essential modifications to address these issues. In agriculture, AI is being utilized to grow healthier crops. It assists with pest management and keeps an eye on soil and environmental conditions. Additionally, it helps analyze data for farmers and improves various tasks related to food supply chain management. For example, figuring out the right time to plant can be tough. Thankfully, AI offers guidance by recommending the best seeds for weather conditions. It even provides valuable weather forecasts. AI-driven solutions will accelerate product launches, enhance crop quality, and help farmers increase yield with fewer resources. AI assists in understanding soil properties and provides recommendations for nutrients to improve soil quality. It also helps farmers determine the best time to plant seeds. Intelligent machinery can identify the optimal planting depth and seed spacing. AI and Machine Learning (ML) are transforming the agricultural sector. Organizations are creating various expertise to assist farmers in monitoring crops and soil health. Advanced AI-based tools such as hyperspectral imaging and 3D laser scanning provide more accurate and comprehensive data for analyzing crop health (Javaid et al. 2023). Over the past two centuries, agricultural productivity has surged with the advent of tractors and harvesters. Currently, artificial intelligence (AI) holds the potential to further increase production affordably and innovatively. This is because, in comparison to human analysis, AI-driven forecasts and modeling can analyze aspects influencing agricultural and food production on a far bigger scale with exceptional precision. The agri-food supply chain is evolving due to global population growth and climate changes. By leveraging vast amounts of data, it is now possible to optimize food production and supply by accurately assessing produce yields and end-user demands (Monteiro and Barata 2021).

AI is altering the food supply chain. It goes from farm to fork, making it safer and more efficient. This change is important for the economy, food safety, and public health. Recently, advances in machine learning have been used to boost food production. They also help with preservation, energy management, and packaging labels in stores. Some techniques play a big role in this area, including Fully Evolutionary System and Extended Temporal Recollections. There are also Sensational Networks that repeat frequently and encode automatically. Attention Mechanisms come into play, too, as well as Latent Variable extraction and grouping methods, and Domain Adaptation. All these deep neural architectures are key for improving the food supply process (Kollia et al. 2021). The implementation of artificial intelligence (AI) in food supply chains (FSC) can significantly enhance transparency and traceability, thereby addressing key concerns such as food safety, quality, and waste. Despite this potential, the existing literature on the implementation of technology in FSC is still in its nascent steps (Dora et al. 2022). As artificial intelligence (AI) rapidly evolves, its concepts are continually shifting. The classification of AI now depends on its development stages, which can be divided into three classifications: analytical, human-inspired, and humanized. Additionally, AI can be classified based on task type (such as numerical data analysis), its form (virtual or actual), the extent of automation and cognitive modeling (basic, advanced, autonomous), and its application functions (descriptive, predictive, prescriptive). Future technological advancements may lead to new forms of AI.

The entire food supply and value chain can be integrated vertically using many functions artificial intelligence provides. Artificial Intelligence (AI) works with other technologies like Machine Learning, Big Data Mining, Internet of Services, Industrial Robots, Agribionts, Digital Platforms, Sensors and Drones, Nanotechnology, Driverless Vehicles, and Machinery. Robotics, drones, and smart machines all interact with these technologies. Each technology serves specific roles at different stages of the supply chain. Still, several obstacles persist. There are technological hurdles, social challenges, and financial issues that hinder the prosperous implementation of AI in this field (Baciuliene et al. 2023). Automation in the food industry depends profoundly on AI, deep learning (DL) algorithms, and machine learning (ML). Implementing AI-based systems can notably enhance the productivity of food production and delivery techniques, enhancing overall operational performance. This article will explore the various AI applications in the food sector, demonstrating how these technologies can lead to significant cost reductions and optimal utilization of resources by minimizing human error. Restaurants, coffee shops, online meal delivery services, lodging facilities, and food outlets can all leverage artificial intelligence and data science to increase productivity and accurately predict sales using various algorithms. AI can substantially enhance attributes such as packaging, shelf life, menu combinations, and food safety. Moreover, implementing a more transparent supply chain management technique would further enhance food safety. The future of the food industry is increasingly relying on advancements like robotic farming, smart farming, and the use of drones (Kumar et al. 2021). The agri-food sector is up against a tough challenge. By 2050, the global population will be more than 9 billion, which means we will need a 70% boost in agricultural and food production to satisfy this rising demand. However, reaching this goal isn't easy. There are limited resources, the effects of climate change, the ongoing impacts of the COVID-19 pandemic, and tough conditions to consider. This situation makes computer tools and forecasting techniques crucial. AI and ML are becoming vital assets in the food and agriculture sectors. Food and agriculture have a significant role in our lives. They are essential for supporting the global population. To get products to consumers, primary agricultural goods go through several stages within a complex supply chain. This process involves preproduction, production, processing, and distribution. Yet, these industries face many hurdles—like growing populations, climate shifts, tech progress, and dwindling natural resources such as water. It is critical to tackle these challenges head-on if we want a sustainable and efficient food system moving forward. Digital technologies should be quickly implemented across various parts of the agricultural supply chain. Some of these technologies involve remote satellite data, automated farm machinery, sensors, machine learning, and artificial intelligence. These tools will enhance crop monitoring and improve systems that track water usage in food production overall (Ben-Ayed and Hanana 2021). Contemporary agriculture and food systems have some goals. They aim to boost output productivity. Also, they focus on operational effectiveness. Resilience to climate change is crucial too. We must think about sustainability for future generations. Because of all this, a technological revolution is needed now more than ever. To meet the challenges of this new way of thinking, artificial intelligence (AI) steps in. AI is a recent advancement in computer science. Understanding how significant AI is and how it can be applied in the agriculture and food sectors might be key. This could play a vital role in ensuring global food security moving forward (Chamara et al. 2020).

FOOD REDISTRIBUTION SYSTEMS POWERED BY AI

AI-powered food reorganization has the capability to decrease food waste and uncertainty by improving food access for underprivileged groups and at-risk people. Moreover, applying artificial intelligence could bring about the following benefits: By matching food donors with organizations that assist people suffering from food poverty, artificial intelligence can greatly cut food waste. AI can help lower waste and strengthen the effectiveness of food distribution by guaranteeing that food donations go to the areas with the most need and in the correct amounts (Sonwani et al. 2021). By improving food access, cost, and stability, AI-powered food reorganization can notably assist underprivileged groups, including the homeless, unemployed, and financially unstable households. By means of resource allocation optimization and identification of high-need locations, artificial intelligence guarantees that food donations reach people at risk (UN Transforming our world 2015). Food banks and nonprofit organizations can examine vast databases on food contributions, population densities, areas, regularity of contribution, and distribution patterns using machine learning (ML) techniques. This approach helps pinpoint the regions most impacted by food hunger and forecast supply and demand. Furthermore, machine learning can identify the nutritional preferences of receivers, which enables the effective and efficient strategic implementation of distribution methods, volumes, and locations (Morgenstern et al. 2021).

Promising results have been seen when using artificial intelligence (AI) to improve waste management and sustainable economic activities. Waste reduction is a vital part of a sustainable economy. AI technology can help spot these opportunities. It can look at new data from different sources, give fast results, and change as needed. This helps governments make smart choices based on current and precise information. In solid waste management (SWM), artificial intelligence also can increase quality, safety, and efficiency. By using artificial intelligence, automation reduces expenses while raising scalability and output. Whether in waste management, facilities, energy, safety, health, or fortification of the atmosphere and social security, the increasing global use of AI algorithms for optimizing SWM reveals how successful these technologies are in addressing challenging issues. Furthermore, especially in the industrial industry, artificial intelligence might help compile data for sustainable goals. It simplifies monitoring and decision-making by using information analysis on items and procedures. Including artificial intelligence can improve resource efficiency in sectors including transportation, sorting, recycling, and collecting as well as waste management processes. Using artificial intelligence to support sustainable economic projects has several benefits since it increases resource value by prolonging product and component life and improving the efficacy of resources. By means of waste reduction and resource enhancement, it also facilitates the attainment of social, economic, and environmental sustainability targets.

Backing sustainable economic projects using artificial intelligence has many advantages. It can first help resources be more effective. It also facilitates the extension of product and part lifetime. Better resource value results from this. Furthermore, artificial intelligence greatly helps in attaining sustainability targets by reducing waste and raising overall efficiency. These objectives cover social and environmental issues as well as economic ones. AI has great application in sectors including transportation, food processing, and agriculture. For wise management of fisheries, for example, data mining exists. Furthermore, in sectors powered by artificial intelligence, drying techniques can be enhanced and energy savings achieved. Then there is crop age estimation based on picture analysis. Integrating artificial intelligence into sustainable projects, supply networks, and food manufacturing looks to have bright results. Over a broad spectrum of businesses, artificial intelligence certainly has the power to improve sustainability, decrease waste, and raise efficiency. Still, it's crucial to make sure the benefits of artificial intelligence exceed the accompanying infrastructure requirements. That requires close inspection of both. Backing sustainable economic projects using artificial intelligence has many advantages. It can first help resources be more effective. It also facilitates the extension of product and part lifetime. Better resource value results from this. Furthermore, artificial intelligence greatly helps in attaining sustainability targets by reducing waste and raising general efficiency. These objectives concern social and environmental concerns as well as the state of the economy.

AI has great applications in sectors including transportation, food processing, and agriculture. For the wise management of fisheries, for example, data mining exists. Furthermore, in sectors powered by artificial intelligence, drying techniques can be enhanced and energy savings achieved. Then there is a crop age estimate based on picture analysis. Integrating artificial intelligence into sustainable projects, supply networks, and food manufacturing looks to have really bright results. In many different sectors, artificial intelligence could improve productivity, lower waste, and raise sustainability. Still, it's crucial to guarantee that the advantages of artificial intelligence outweigh the demand of the requisite infrastructure. That requires a close examination of the environmental as well as financial aspects involved.

CONCLUSIONS

One major issue worldwide is food waste. It has different effects on society, the economy, and the environment. Fortunately, one can solve this problem by advocating waste reduction and improved resource economy. Here, a sustainable economic strategy can really make a difference. AI has lately become increasingly popular in the food sector. It's meant to help with food waste and increase environmental initiatives. Food output is improved, and supply chains are simplified in part by artificial intelligence. It also encourages cutting waste and recycling. It also makes getting extra food to people in need much simpler. Utilizing artificial intelligence technologies will help us maximize resources. This lessens our environmental impact and advances a more equitable and sustainable food chain. Projects using artificial intelligence to assist sustainability have several benefits, including improved decision-making, longer product shelf life, and higher energy efficiency. Constant investment in artificial intelligence technology is crucial if we are to fully realize the possibilities of a sustainable economy and properly address global food waste.

REFERENCES

Ahmed, A.A.A. and Asadullah, A. 2020. Artificial intelligence and machine learning in waste management and recycling. Engineering International. 8(1): 43–52.

Amicarelli, V., Aluculesei, A.C., Lagioia, G., Pamfilie, R. and Bux, C. 2022. How to manage and minimize food waste in the hotel industry: An exploratory research. International Journal of Culture, Tourism and Hospitality Research. 16(1): 152–167.

Andeobu, L., Wibowo, S. and Grandhi, S. 2022. Artificial intelligence applications for sustainable solid waste management practices in Australia: A systematic review. Science of The Total Environment. 834: 155389.

Aniza, R., Chen, W.H., Pétrissans, A., Hoang, A.T., Ashokkumar, V. and Pétrissans, M. 2023. A review of biowaste remediation and valorization for environmental sustainability: Artificial intelligence approach. Environmental Pollution. 324: 121363.

Bačiulienė, V., Bilan, Y., Navickas, V. and Civín, L. 2023. The aspects of artificial intelligence in different phases of the food value and supply chain. Foods. 12(8): 1654.

Batabyal, D. and Das, D.K. 2022. Socio-economic development through tourism: An investigative study for the Himalayan State Sikkim, India. *In*: Jana, N.C., Singh, A., Singh, R.B. (eds). Livelihood Enhancement Through Agriculture, Tourism and Health. Advances in Geographical and Environmental Sciences. Springer, Singapore. https://doi.org/10.1007/978-981-16-7310-8_12.

Batabyal, D. 2010. Implication of a scientific destination study in tourism product management: A case study in Sikkim. *In*: Article Presented and Published in a National Conference Science and Technology Applications in Tourism Sector Focusing on Uttarakhand Opportunities.

Ben-Ayed, R. and Hanana, M. 2021. Artificial intelligence to improve the food and agriculture sector. Journal of Food Quality. 2021(1): 5584754.

Berman, S.D. 2014. Hospitality Directions US: Our Updated Lodging Outlook [wwwdocument]. Price Waterhouse Coopers. http://www.pwc.com/en_US/us/asset-management/hospitality-leisure/publications/assets/pwc-hospitality-directions-lodging-outlook.pdf.

Calvo-Porral, C., Medín, A.F. and Losada-López, C. 2017. Can marketing help in tackling food waste? Proposals in developed countries. Journal of Food Products Marketing. 23(1): 42–60.

Canton, H. 2021. Food and agriculture organization of the United Nations—FAO. pp. 297–305. *In*: edited By Europa Publications. The Europa Directory of International Organizations 2021, 23rd Edition. Routledge.

Chamara, R.M.S.R., Senevirathne, S.M.P., Samarasinghe, S.A.I.L.N., Premasiri, M.W.R.C., Sandaruwani, K.H.C., Dissanayake, D.M.N.N., et al. 2020. Role of artificial intelligence in achieving global food security: A promising technology for future. Sri Lanka Journal of Food and Agriculture. 6(2).

Charney, M. 2009. Food routes network and the local food movement. Journal of Agricultural and Food Information. 10(3): 173–181.

Delgado, L., Schuster, M. and Torero, M. 2017. Reality of food losses: A new measurement methodology.

Dora, M., Kumar, A., Mangla, S.K., Pant, A. and Kamal, M.M. 2022. Critical success factors influencing artificial intelligence adoption in food supply chains. International Journal of Production Research. 60(14): 4621–4640.

Ellen Mac Arthur (Foundation). 2015. Growth within: A Circular Economy Vision for a Competitive Europe. Ellen MacArthur Foundation: Isle of Wight, UK.

Filimonau, V. and Delysia, A. 2019. Food waste management in hospitality operations: A critical review. Tourism Management. 71: 234–245.

Frayne, B., Pendleton, W., Crush, J., Acquah, B., Battersby-Lennard, J., Bras, E., et al. 2010. The state of urban food insecurity in southern Africa. Urban Food Security Series. 2: 1–54.

Gustavsson, J., Cederberg, C., Sonesson, U., Van Otterdijk, R. and Meybeck, A. 2011. Global food losses and food waste.

Ihsanullah, I., Alam, G., Jamal, A. and Shaik, F. 2022. Recent advances in applications of artificial intelligence in solid waste management: A review. Chemosphere. 309: 136631.

Isabelle, D., 2014. One third of all food wasted. FAO Liaison Office in Brussels (Accessed 11.03.15). http://www.unric.org/en/food-waste/27133-one-third-of-all-food-wasted.

Jamaludin, H., Elmaky, H.S.E. and Sulaiman, S. 2022. The future of food waste: Application of circular economy. Energy Nexus. 7: 100098.

Javaid, M., Haleem, A., Khan, I.H. and Suman, R. 2023. Understanding the potential applications of Artificial Intelligence in Agriculture Sector. Advanced Agrochem. 2(1): 15–30.

Kollia, I., Stevenson, J. and Kollias, S. 2021. AI-enabled efficient and safe food supply chain. Electronics. 10(11): 1223.

Kumar, I., Rawat, J., Mohd, N. and Husain, S. 2021. Opportunities of artificial intelligence and machine learning in the food industry. Journal of Food Quality. 2021(1): 4535567.

Kumar, M., Raut, R.D., Jagtap, S. and Choubey, V.K. 2022. Circular economy adoption challenges in the food supply chain for sustainable development. Business Strategy Environment. 32: 1334–1356.

Mboli, J.S., Thakker, D. and Mishra, J.L. 2022. An Internet of Things-enabled decision support system for circular economy business model. Software: Practice and Experience. 52(3): 772–787.

McKinsey & Company. 2019. How AI Can Unlock a $127 B Opportunity by Reducing Food Waste; McKinsey & Company: Atlanta, GA, USA.

Mganga, P.P., Syafrudin, S. and Amirudin, A. 2021. Students' Awareness on Food Waste Problems and their Behaviour towards Food Wastage: A Case Study of Diponegoro University (UNDIP) Tembalang Campus. Master's Thesis, School of Postgraduate Studies, Diponegoro University, Kota Semarang, Indonesia.

Monteiro, J. and Barata, J. 2021. Artificial intelligence in extended agri-food supply chain: A short review based on bibliometric analysis. Procedia Computer Science. 192: 3020–3029.

Morgenstern, J.D., Rosella, L.C., Costa, A.P., de Souza, R.J. and Anderson, L.N. 2021. Perspective: Big data and machine learning could help advance nutritional epidemiology. Advances in Nutrition. 12: 621–631.

Munesue, Y., Masui, T. and Fushima, T. 2015. The effects of reducing food losses and food waste on global food insecurity, natural resources, and greenhouse gas emissions. Environmental Economics and Policy Studies. 17: 43–77.

Ndondo, J.T.K. 2023. Review of the food and agriculture organisation (FAO) strategic priorities on food safety 2023. *In*: Food Safety - New Insights. IntechOpen. DOI: 10.5772/intechopen.112132

Onyeaka, H., Tamasiga, P., Nwauzoma, U.M., Miri, T., Juliet, U.C., Nwaiwu, O., et al. 2023. Using artificial intelligence to tackle food waste and enhance the circular economy: Maximising resource efficiency and Minimising environmental impact: A review. Sustainability. 15(13): 10482.

Pirani, S.I. and Arafat, H.A. 2016. Reduction of food waste generation in the hospitality industry. Journal of Cleaner Production. 132: 129–145.

SaMurAI®-Sorting Robot-MachineX. 2025. Machinex. https://www.machinexrecycling.com/sorting/equipment/samurai-sorting-robot/

Sharma, S., Gahlawat, V.K., Rahul, K., Mor, R.S. and Malik, M. 2021. Sustainable innovations in the food industry through artificial intelligence and big data analytics. Logistics. 5(4): 66.

Sonwani, E., Bansal, U., Alroobaea, R., Baqasah, A.M. and Hedabou, M. 2021. An artificial intelligence approach toward food spoilage detection and analysis. Frontiers in Public Health. 9: 816226.

Tamasiga, P., Miri, T., Onyeaka, H. and Hart, A. 2022. Food waste and circular economy: Challenges and opportunities. Sustainability. 14(16): 9896.

Thakar, M. 2019. Award-winning AI Technology Helps IHG Hotels Track, Measure and Reduce Food Waste. https://www.ihgplc.com/news-and-media/news-releases/2019/award-winning-ai-technology-helps-ihg-hotels-track-measure-and-reduce-food-waste

Thi, N.B.D., Kumar, G. and Lin, C. 2015. An overview of food waste management in developing countries: Current status and future perspective. Journal of Environmental Management. 157: 220–229. https://doi.org/10.1016/j.jenvman.2015.04.022

UN Transforming our world. 2015. The 2030 Agenda for Sustainable Development. *In*: Division for Sustainable Development Goals. Springer: NewYork, USA.

UNEP Food Waste Index Report. 2024. https://www.civilsdaily.com/news/unep-food-waste-index-report-2024/

United Nations. UNEP Food Waste Index Report. 2021. Available online: http://www.unep.org/resources/report/unep-foodwaste-index-report-2021 (accessed on 27 March 2023).

Van der Werf, P. and Gilliland, J.A. 2017. A systematic review of food losses and food waste generation in developed countries. *In*: Proceedings of the Institution of Civil Engineers-Waste and Resource Management. 170(2): 66–77. Thomas Telford Ltd.

Van Wynsberghe, A. 2021. Sustainable AI. AI for sustainability and the sustainability of AI. AI and Ethics. 1(3): 213–218.

Vinuesa, R., Aszizpour, H., Leite, I., Balaam, M., Dignum, V., Domisch, S., et al. 2020. The role of artificial intelligence in achieving the Sustainable Development Goals. Nature Communications. 11(1): 1–10.

Chapter 3

Smart Maintenance: AI and IoT Driven Asset Care in Hospitality

Sunil Sharma*[1], Vijendra Kumar Maurya[2], Prasun Chakrabarti[3], Arvind Singh Rathore[4], Sandip Das[5], Bhupendra Soni[6] and Prashant Sharma[7]

[1]Department of CES, Techno NJR Institute of Technology, Udaipur, India

[2]Department of ECE, GITS, Udaipur, India

[3]Pro-VC, SPSU, Udaipur, India

[4]Department of Computing and Informatics, SPSU, Udaipur, India

[5]Department of CSE, Brainware University, W.B., India

[6]Department of ECE, ACEIT, Jaipur, India

[7]Department of CSE, VGU, Jaipur, India

INTRODUCTION

Hospitality environments today require more than traditional maintenance systems because these approaches generally demand excessive manual work and delayed responses (Ali and Amin 2014). Hotel asset care undergoes a complete transformation because of artificial intelligence (AI), which brings disruptive changes to the industry. As part of artificial intelligence, asset care maintenance has transitioned from randomly fixed events to advanced proactive methods with the utilization of machine learning predictive analytics together with Internet

*For Correspondence: ersharma.sunil@gmail.com

DOI: 10.1201/9781003540083-3

of Things (IoT) technology. The authors (Takuli et al. 2025) explored robotic technology growth within hospitality along with its operational benefits for visitors. Artificial Intelligence allows hospitality companies to enhance their maintenance scheduling to save expenses and forecast equipment breakdowns before failures occur. The strategic approach delivers visitor satisfaction through smooth experiences, lengthens key assets, and maintains their reliability. AI-driven asset care presents itself as a transformative operational innovation that changes the hospitality industry through superior sustainability, operational efficiency, and comprehensive excellence. The sector operates at heightened speed in its adoption of digital transformation across its operations.

FUNDAMENTAL ROLE OF AI IN HOSPITALITY

This feature enables companies in the hospitality industry to switch from reactive to proactive maintenance methods. Many contemporary authors have discussed the main characteristics and uses of the Internet of Things (IoT), highlighting how it can be used to create smart, connected environments, including those in the hospitality industry (Sharma et al. 2023). Another important benefit of AI-driven asset care is cost reduction. Artificial Intelligence reduces wasteful spending by optimizing maintenance schedules and resources. Conventional maintenance procedures often entail frequent inspections and fixes regardless of real need, wasting resources and time. Conversely, artificial intelligence (AI) can precisely determine when and where maintenance is needed, which lowers labor costs and the frequency of meaningless repairs. AI-driven energy management systems also have the ability to monitor and modify energy consumption in real time, which can result in substantial savings. (Buhalis and Amaranggana 2015) investigated how individualized services are used by smart tourism destinations to improve the traveler experience via ICT innovations. For example, occupancy patterns can be learned by smart HVAC systems, which can then adjust temperatures to save up to 15% on energy. In the hotel industry, Figure 3.1 illustrates the basic functions of AI.

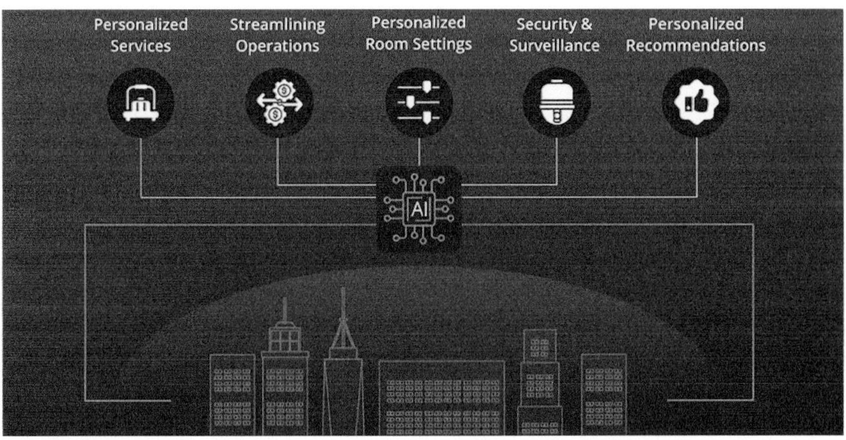

Figure 3.1 Fundamental role of AI in the hospitality industry (Ted A. Fernandez 2024).

Enhancing asset care through AI considerably improves the overall visitor experience. In the hospitality sector, customer satisfaction is of utmost importance, and any interruption in service can have a detrimental effect on the guest's opinion. In their study, (Buhalis and Leung 2018) explored how smart hospitality technologies are integrated, highlighting the significance of interconnectivity and interoperability in building an ecosystem. Improved operational reliability reduces the likelihood that visitors will experience problems like broken air conditioning or unannounced maintenance noise while visiting. With the ability to learn preferences, AI can customize the guest experience by modifying services in real time. For example, it can recommend customized amenities or set the ideal room temperature when a guest checks in.

PREDICTIVE ANALYTICS, MACHINE LEARNING, AND THE INTERNET OF THINGS (IoT) IN HOSPITALITY

(Choi and Kandampully 2019) identified important areas for improvement after analyzing the effects of technolozsgy on service interactions and customer service experiences in the hospitality industry. Together, these technologies improve cost-effectiveness, operational efficiency, and guest satisfaction by facilitating more informed data-driven decision-making and proactive management. In order to predict future events, predictive analytics analyzes both historical and current data.

Predictive analytics is a useful tool in the hospitality industry for better inventory management, staffing optimization, and maintenance forecasting. (Connolly and Olsen 2017) put forth a framework for environmental scanning that focuses on navigating technological changes for strategic planning in the hospitality industry. This keeps crucial systems running constantly and minimizes unscheduled downtime. Furthermore, hotels can improve efficiency and service quality by adjusting staffing and resources based on the predicted demand from guests, which is made possible by predictive analytics. Figure 3.2 illustrates the use of predictive analytics in the hotel sector.

MACHINE LEARNING

Algorithms that aren't explicitly programmed learn from data and become more sophisticated over time as part of the machine learning subset of artificial intelligence Through the analysis of historical behaviors and preferences, machine learning can be applied to the hospitality industry to personalize guest experiences. In order to improve each visitor experience, machine learning algorithms, for instance, can recommend customized dining options, room settings, or marketing messages. (Evans 2021) revealed how to incorporate technology into strategic planning, offering valuable insights into strategic management for the travel and hospitality sector in 2021. The role of predictive machine learning in the hospitality industry is depicted in Figure 3.3.

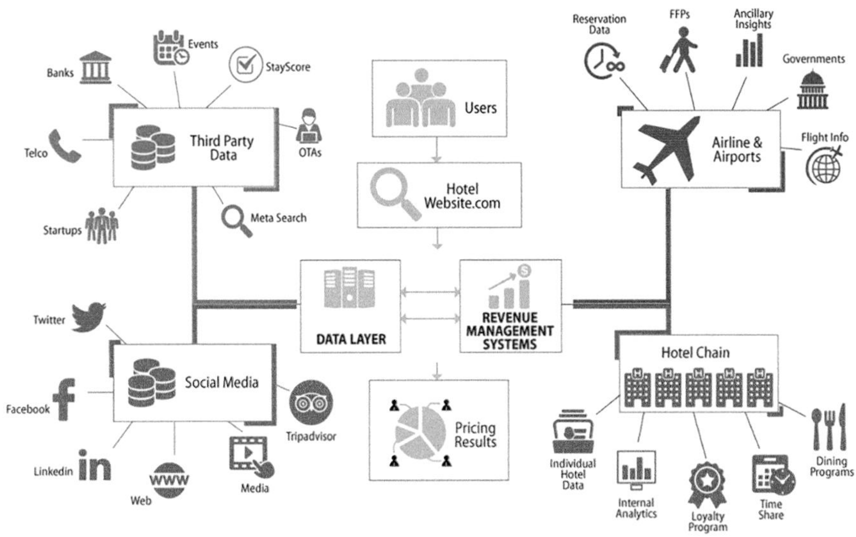

Figure 3.2 Predictive analytics in hospitality (Mark Ross-Smith 2016).

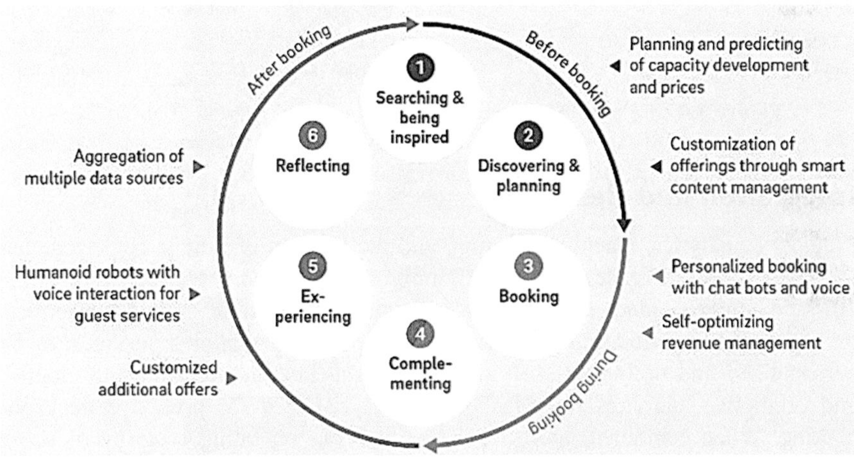

Figure 3.3 Machine learning in hospitality (Sharma et al. 2022).

INTERNET OF THINGS (IoT)

The term Internet of Things (IoT) describes a network of devices that are connected to one another and exchange data. IoT devices in the hospitality industry can include anything from linked kitchen appliances and security systems to smart lighting controls and thermostats. Increased productivity and guest comfort are the results of real-time monitoring and control of hotel operations made possible by IoT. For example, occupancy-based smart HVAC systems can automatically adjust room temperature to ensure maximum comfort while consuming less energy.

(Gretzel et al. 2015) and others laid the groundwork for smart tourism in 2015 by examining the ways in which digital technologies are changing travel experiences. IoT-enabled appliances can notify workers of maintenance requirements before problems get worse, guaranteeing a seamless and continuous visitor experience. Furthermore, real-time monitoring and alerts for possible security breaches are two more ways that IoT can improve security. The Internet of Things (IoT), machine learning, and predictive analytics are compared in the context of their applications in the hospitality sector in the following Table 3.1, which also includes important data and advantages for each technology. The function of IoT in the hospitality sector is depicted in Figure 3.4.

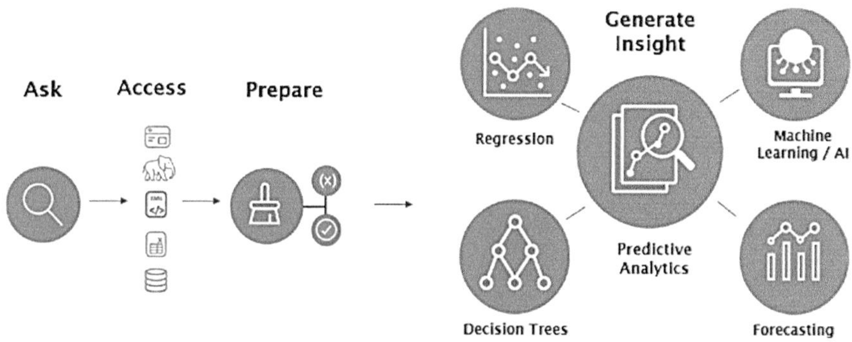

Figure 3.4 Role of IoT in the hospitality industry (Sharma et al. 2023).

Integration and Benefits

Predictive analytics, machine learning, and the Internet of Things come together to form a potent ecosystem that greatly improves the hospitality sector (Guttentag 2015). A study conducted on Airbnb's disruptive innovation outlined its effects on the traditional lodging industry. Machine learning allows services to be personalized and optimized, IoT guarantees real-time connectivity and control, and predictive analytics provides the insight required for proactive decision-making. When combined, these technologies lower operating costs by reducing energy consumption and maintenance expenses, boost productivity by allocating resources more effectively, and enhance the visitor experience with seamless operations and individualized services. For instance, by employing these technologies, a hotel can anticipate peak times, adjust staffing levels, tailor guest experiences according to preferences discovered from previous stays, and use IoT monitoring to ensure every system is operating at peak efficiency. (Choi and Kandampully 2019) highlighted potential advantages and difficulties of AI in travel, tourism, and hospitality. Smarter, more individualized, and efficient operations are made possible by IoT, machine learning, and predictive analytics, which are revolutionizing the hospitality sector. By integrating them, hotels are better positioned to prosper in a market that is becoming more competitive while also increasing operational efficiency, cutting costs, and improving the overall

Table 3.1 Applications and key benefits of different technologies in the hospitality industry

Technology	Predictive Analytics	Machine Learning	Internet of Things (IoT)
Function	Analyzes historical and real-time data to forecast future events.	Algorithms learn from data to improve over time and make decisions.	A network of interconnected devices that collect and exchange data.
Key Benefits	Anticipates maintenance needs Optimizes staffing levels Improves inventory management	Personalizes guest experiences Optimizes pricing strategies Enhances marketing efforts	Real-time monitoring and control Enhances operational efficiency Improves guest comfort and security
Maintenance Efficiency	Reduction in emergency repairs by 25% Optimized maintenance schedules reduce downtime by 20%	Efficient resource allocation reduces unnecessary maintenance by 30%	Continuous monitoring ensures timely maintenance actions Real-time alerts reduce downtime by 15%
Guest Experience	Improved service predictability Enhanced satisfaction through proactive service	Personalized recommendations increase guest satisfaction by 20%	Automated adjustments (e.g. HVAC, lighting) based on occupancy Improved comfort and convenience
Operational Cost Savings	Reduces unplanned maintenance costs by 20% Optimizes resource usage	Reduces labor costs through efficient task allocation Increases revenue through dynamic pricing by 15%	Reduces energy consumption by 15% Lowers operational costs through automation
Energy Efficiency	Identifies energy saving opportunities Forecasts energy needs accurately	Optimizes energy use through learning patterns Predicts peak usage times to reduce costs	Real-time energy management reduces consumption by 15% Smart systems adjust energy use dynamically
Asset Lifespan	Extends asset lifespan by predicting failures and scheduling maintenance	Improves asset management through continuous learning Prolongs asset life by optimizing usage patterns	Prolongs equipment life through continuous monitoring Preventive maintenance reduces wear and tear by 20%
Statistics	20% reduction in maintenance costs 25% increase in asset reliability	20% increase in guest satisfaction 15% increase in revenue through optimized pricing	15% reduction in energy consumption 20% increase in operational efficiency

guest experience. (Ivanov and Webster 2020) presented a thorough analysis of current trends in the development of a conceptual framework for the application of robots, artificial intelligence, and service automation in the tourism industry in 2020. AI integration changes the operational environment of the hotel sector. Businesses can provide better, continuous guest experiences, increase efficiency, and save operating expenses with its help. These advancements raise the bar for industry excellence by increasing competitiveness, long-term sustainability, and customer loyalty. Philip Jones emphasized that technology plays a crucial role in achieving sustainability goals. (Jones et al. 2016) investigated sustainability practices in the hospitality industry.

Practical Applications of AI-Driven Maintenance in Hospitality

AI-driven maintenance systems collect real-time data that produces accurate equipment failure predictions, allowing better maintenance procedures and longer-lasting vital assets. AI-based technologies allow hospitality companies to reach new peaks of operational performance and reliability when properly implemented. This section presents actual applications of AI-driven maintenance, which include:

Equipment Failure Prediction

The identification of equipment failures in advance serves as a function of predictive analytics within AI-driven maintenance systems. Artificial intelligence algorithms process information from Internet of Things devices and sensors about equipment performance aspects like vibration patterns, temperature fluctuations, and energy consumption. An HVAC system with built-in sensors within hotel facilities will detect strange noises and abnormal temperature variations. The AI system utilizes collected data to predict future equipment failures, enabling maintenance workers to perform repairs before breakdowns happen, thus maintaining guest satisfaction.

Making Maintenance Schedules More Effective

Current maintenance practices operate from predetermined calendars that lead to either excessive or inadequate servicing of assets. The optimization of maintenance schedules through AI uses authentic data regarding equipment condition and usage rates. An AI system can analyze live data of water quality chemicals alongside pump performance throughout a resort that contains multiple swimming pools.

Increasing the Lifespan of Important Assets

Through optimal asset maintenance under AI-driven control, the likelihood of failures decreases while assets achieve extended operational lifespans. AI can, for instance, monitor wear and tear on the washing and drying machines in a hotel's laundry room. AI systems can make recommendations on maintenance tasks like part replacements or lubrication by analyzing data on usage frequency, load sizes, and operational efficiency. With a longer operational life and fewer expensive replacements required, this proactive approach helps keep equipment in top condition.

Real-World Examples

Hilton Hotels

Hilton optimizes its kitchen equipment maintenance schedules with the help of artificial intelligence. The AI system plans maintenance for off-peak hours to guarantee that equipment is always in working order during peak service hours. It does this by evaluating data from connected ovens, refrigerators, and dishwashers. Cooking operations are now more efficient, and emergency repairs are down as a result.

Hyatt Hotels

Hyatt uses artificial intelligence (AI) to track the operation of its elevator systems. Motor temperature and door cycle counts are two metrics that are monitored by sensors. The AI system evaluates this data to forecast when components are likely to fail, facilitating timely maintenance and lowering the possibility of elevator downtime, which is crucial for maintaining guest satisfaction in high-rise hotels.

Marriott International

In a number of its locations, Marriott has begun using AI-driven maintenance. The company monitors HVAC systems in order to anticipate and prevent failures through the use of IoT sensors and predictive analytics (Kar 2020) examined the variables influencing users' happiness with mobile payments and created a model of digital service usage satisfaction that is applicable to the hospitality industry. Because there have been fewer disruptions, this initiative has improved guest satisfaction and significantly reduced maintenance costs (Kasavana 2017).

CASE STUDIES OF AI IMPLEMENTATION IN HOSPITALITY

The paper by (Krizaj and Cvelbar 2019) provided an analysis of research in tourism innovation, which revealed essential behavioral patterns alongside technological advancement directions. Observational findings about AI implementation emerge from case studies, which demonstrate high-tech capabilities for maintenance operations and operational efficiency improvements.

Marriott International: The comfort assurance for guests at Marriott International properties stems from their deployment of AI-driven maintenance throughout various locations, which specifically targets HVAC system supervision. By utilizing Internet of Things sensors on HVAC units, Marriott acquires current details about temperature levels, humidity rates, and system operational performance. Using this information, AI programs analyze probable system failures as well as required maintenance tasks beforehand.

Results

Predictive maintenance through Marriott resulted in a 25% decline in HVAC system downtime, which secured unbroken comfort for guests throughout their stay. The proactive identification of maintenance problems enables Marriott to reduce expenses by 20%.

Graph 1 represents the impact of AI-driven maintenance on Marriott International's Halifax HVAC systems in 2025.

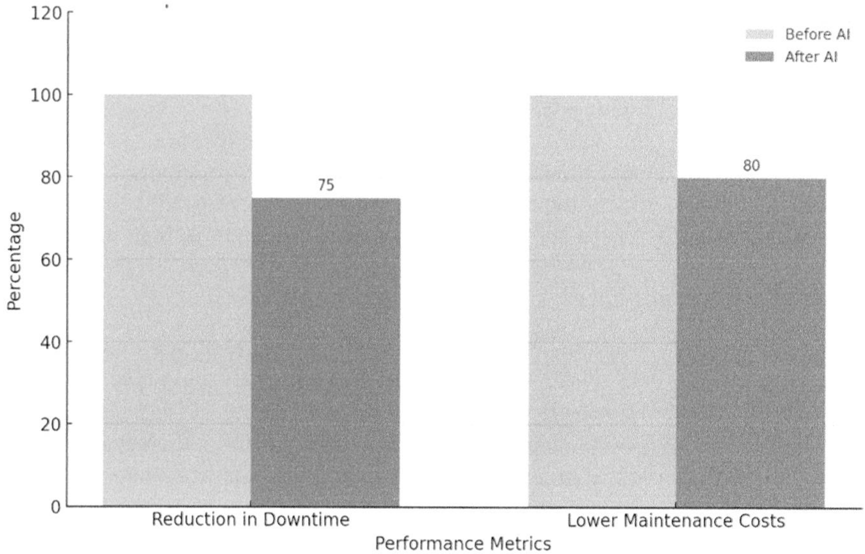

Graph 1 Impact of AI-driven maintenance on Marriott International's HVAC systems.

The impact of AI-driven maintenance on Marriott International's Halifax HVAC systems, showing the reduction in downtime and maintenance costs before and after AI implementation.

Hilton Hotels: Optimizing Kitchen Equipment Maintenance

AI has been used by Hilton Hotels to improve the maintenance plans for its kitchen appliances, including ovens, refrigerators, and dishwashers. Sensors keep an eye on temperature, equipment utilization, and overall performance. By using this data to schedule maintenance during times of low demand, the AI system avoids service interruptions during periods of high demand.

Results

Improved Equipment Uptime: Hilton has increased kitchen equipment uptime by thirty percent, guaranteeing that kitchens run smoothly and effectively during peak hours.

Cost Savings: Because preventive maintenance lowers the need for expensive emergency repairs, optimized maintenance schedules have resulted in a 15% reduction in maintenance costs.

Graph 2 shows the impact of AI-driven maintenance on Hilton Hotels' McLean (2024) kitchen equipment.

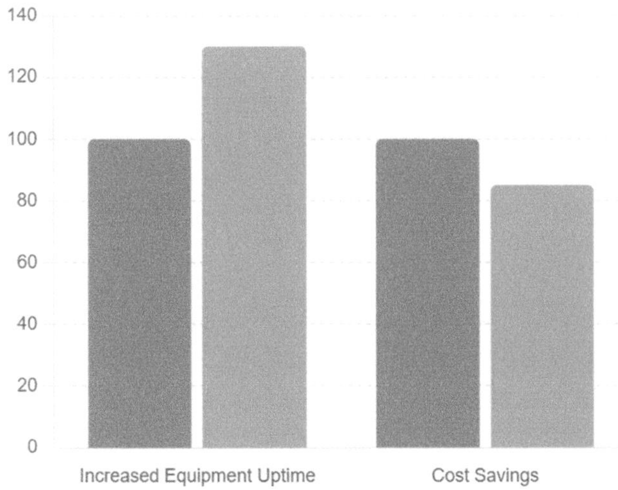

Graph 2 Impact of AI-driven maintenance on Hilton Hotels' kitchen equipment.

Impact of AI-driven maintenance on Hilton Hotels' kitchen equipment, showing the increased equipment uptime and cost savings before and after AI implementation.

Hyatt Hotels: Enhancing Elevator Reliability

AI-driven maintenance has been introduced by Hyatt Hotels for its elevator systems, which are essential for visitor mobility, particularly in high-rise buildings. Door cycles, motor temperature, and usage trends are all monitored by sensors. This data is analyzed by AI to forecast when parts will break and to plan preventive maintenance.

Results

Enhanced Dependability: Up to 25% more maintenance is completed efficiently now that problems are being fixed before they become major system failures, thanks to the predictive maintenance strategy.

Graph 3 presents the impact of AI-driven maintenance on Hyatt Hotels' (2024) elevator systems.

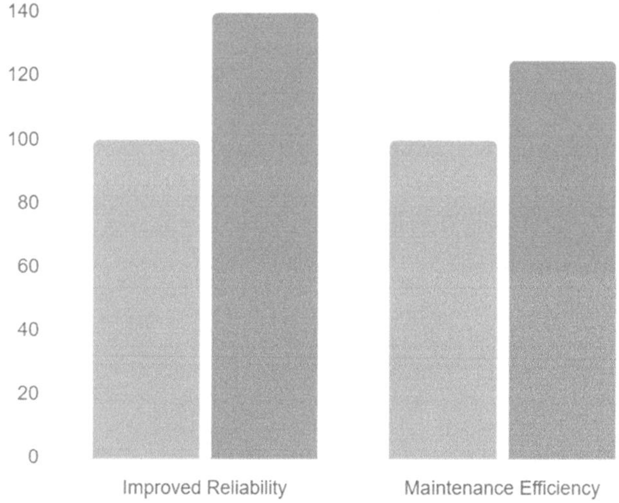

Graph 3 Impact of AI-driven maintenance on Hyatt Hotels' elevator systems.

Graph 3 shows a before-and-after comparison revealing how AI-driven maintenance enhanced elevator system reliability and maintenance efficiency at Hyatt Hotels.

Accor Hotels: Energy Consumption and Sustainability

Accor Hotels has concentrated on cutting down on energy use. Smart lighting HVAC and appliance energy consumption are all tracked by Internet of Things (IoT) sensors.

Results

Energy Savings: By streamlining energy use, supporting sustainability initiatives, and cutting expenses, Accor has achieved energy savings of up to 20 percent.

Enhanced Sustainability: Accor's dedication to sustainability is reflected in the energy consumption reduction, which makes its properties more environmentally friendly and desirable to guests who care about the environment.

Graph 4 presents the impact of AI-driven energy management on Accor Hotels, Saudi Arabia (2024).

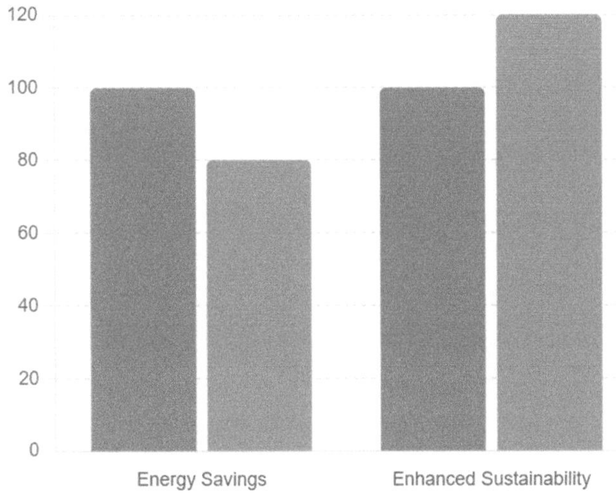

Graph 4 Impact of AI-driven energy management on Accor Hotels.

Graph 4 showing the energy saving and enhanced sustainability before and after the implementation of AI.

Wyndham Hotels and Resorts: Comprehensive Facility Management

AI-driven maintenance has been implemented by Wyndham Hotels and Resorts to manage all aspects of the facility, including plumbing, lighting, and HVAC systems. Predictive and preventive maintenance is made possible by real-time data collection and artificial intelligence (AI) analysis, which offer insights into system operation.

Results

Operational Efficiency: Wyndham claims a thirty percent boost in overall operational efficiency thanks to AI-driven maintenance that makes sure all vital systems are operating at peak performance.

Cost Reductions: Because fewer emergency repairs and routine inspections are required, the application of AI-driven maintenance has resulted in a 25% decrease in total maintenance costs.

Graph 5 presents the impact of AI-driven maintenance on Wyndham Hotels and Resorts, Shanghai (2024).

The impact of AI-driven maintenance on Wyndham Hotels & Resorts, showing the increase in operational efficiency and cost reductions before and after AI implementation.

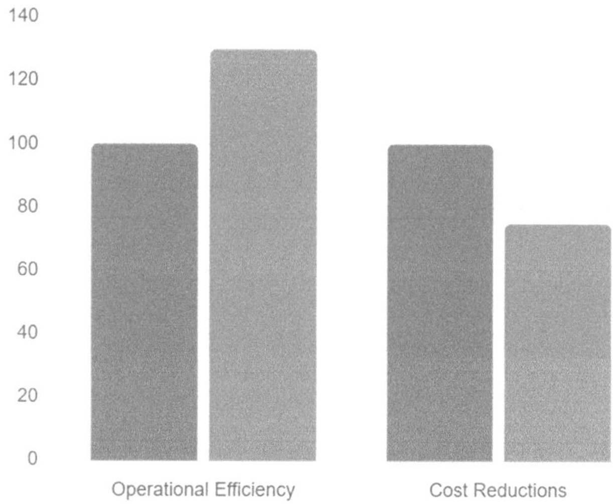

Graph 5 Impact of AI-driven maintenance on Wyndham Hotels & Resorts.

Comprehensive Comparison: Conventional vs. AI-Powered Maintenance

The statistical insights into the performance variations between traditional and AI-powered maintenance in the hospitality industry are presented in Table 3.2 below for a thorough comparison. The following comparative statistics in Table 3.2 indicate that the integration of AI technologies has significantly improved several operational aspects (Lu and Cecil 2016). The speaker delineated AI and robotic systems' capacity to reconstruct the hospitality industry through advanced guest service and operational system optimization.

AI-Powered Systems in Hospitality Maintenance

The predictive power of artificial intelligence systems allows for proactive maintenance initiatives that minimize disruptions and prolong vital asset lifespans. The systems leverage accurate real-time data analysis alongside complex algorithms to forecast potential issues before recommending maintenance actions that protect operational effectiveness. The latest hospitality trends discussed by Rawal et al. (2020, 2022) include IoT, AI, and sustainable practices, which demonstrate their effects on the industry. Multiple essential performance indicators demonstrate that AI-based maintenance exceeds traditional maintenance practices in performance results.

A Rise in the Reliability of Assets

Technology components consisting of IoT sensors and devices work with AI software to provide constant monitoring of asset performance as well as

Table 3.2 Statistical insights between conventional and AI-powered maintenance in the hospitality industry

Aspect	Conventional Maintenance	AI-Powered Maintenance	Improvement
Downtime Reduction	Relies on scheduled checks and reactive repairs. Average downtime per year: 50 hours per equipment unit.	Uses predictive analytics to anticipate failures and schedule maintenance proactively. Average downtime per year: 35 hours per equipment unit.	Reduction in Downtime: 30%
Maintenance Costs	Higher costs due to frequent emergency repairs and routine checks. Average annual maintenance cost: $100,000 per hotel.	Lower costs due to optimized maintenance schedules and fewer emergency repairs. Average annual maintenance cost: $80,000 per hotel.	Cost Savings: 20%
Energy Consumption	Inefficient energy use due to lack of real-time optimization. Average annual energy consumption: 1,200,000 kWh per hotel.	Optimized energy use through real-time monitoring and adjustments (Takuli et al. 2025) Average annual energy consumption: 1,020,000 kWh per hotel.	Energy Savings: 15%
Equipment Lifespan	Shorter equipment lifespan due to reactive maintenance. Average lifespan of HVAC units: 10 years.	Extended equipment lifespan through predictive and preventive maintenance. Average lifespan of HVAC units: 12 years.	Increase in Lifespan: 20%
Operational Efficiency	Reactive approach leads to inefficiencies and higher operational disruptions. Average operational efficiency score: 70 (on a scale of 100).	Proactive maintenance reduces disruptions and increases efficiency. Average operational efficiency score: 85 (on a scale of 100).	Improvement in Operational Efficiency: 21%
Guest Satisfaction	Frequent disruptions due to unexpected equipment failures. Average guest satisfaction score: 80 (on a scale of 100).	Fewer disruptions and smoother operations enhance guest experience. Average guest satisfaction score: 90 (on a scale of 100).	Increase in Guest Satisfaction: 12%

operational states. Forecasting asset failure occurs because these systems process information about temperature variations, vibration patterns, and operating hours to assess failure probabilities. The predictive feature of this system guarantees maximum asset reliability because maintenance technicians can solve issues before equipment breakdowns occur.

Statistical Wisdom

Energy facilities implementing AI-driven maintenance achieve improved asset dependability to the extent of 25%. Furthermore, this data demonstrates how the facility achieves better infrastructure dependability together with reduced unplanned equipment breakdowns in its essential machinery systems.

Statistical Knowledge

The implementation of artificial intelligence systems enhances maintenance operations by 40% in addition to standard practices. Improved efficiency results from decreased downtime, fewer unnecessary maintenance operations, and a strategic approach to maintenance work which shows an increase of 40%. By using such facilities more efficiently, they can distribute maintenance resources in smarter ways to reach both cost reduction and increased output levels.

Practical and Feasible Outcomes of AI-Driven Maintenance in Hospitality

The integration of AI-driven maintenance creates enhanced asset reliability, which produces satisfied guests with continuous services, along with long-term cost-saving benefits among numerous realistic, achievable outcomes for the hospitality industry. The business model achieves improved sustainability, operational efficiency, and better competitiveness through such results.

Increased Dependability of the Assets

When hospitality facilities use AI for maintenance, their critical operational assets become much more dependable. As a proactive measure, this methodology enables maximum operational efficiency of kitchen appliances, elevators, and HVAC systems by minimizing operational downtime.

Results

Predictive maintenance extends the operational life of assets by conducting proactive maintenance beforehand.

Topical Transition

The following section will analyze sales forecasting based on historical data. The need for replacements and capital investment decreases due to this approach.

Maintaining Consistent Performance

Assets will operate reliably if their performance levels stay steady, thus protecting operations from unexpected failures that negatively impact visitor satisfaction. The hospitality industry focuses on client satisfaction above all else, which means maintaining uninterrupted services will lead to lasting high levels of satisfaction. Automatic system maintenance through AI keeps essential operations functional, preventing any inconveniences that could disturb visitors. Standardized HVAC system maintenance provides visitors with dependable temperature controls, along with functioning elevators that deliver safe transportation services.

Results

Guest satisfaction remains unburdened since maintenance tasks and equipment failures produce no disruptive events. The overall satisfaction of guests increases when there is no maintenance disruption during their stay, thus generating more chances of favorable customer feedback and repeat business.

Proactive Service

The visitor experience receives personalized treatment through AI-driven systems because these systems have the ability to forecast needs and preferences. AI-driven technology adjusts room control systems through monitored guest conduct and individual tastes. The implementation of AI-driven maintenance provides organizations with a lasting opportunity to minimize their costs.

RESULTS

The replacement plans from efficient maintenance strategies lower both maintenance repairs and labor expenses. Predictive maintenance operates proactively to lower the necessity of unexpected emergency repair operations.

The energy management systems enhanced by artificial intelligence greatly decrease operational energy consumption to lower environmental impact, along with utility expenses. The application of AI-driven maintenance for business operations allows hotels to decrease their expenses, thereby promoting environmentally friendly business practices. The combination of lower energy consumption and asset longevity will let hotels satisfy sustainability requirements and draw sustainability-minded customers to their establishments.

CONCLUSION

The use of AI-powered maintenance systems reduces equipment breakdown periods by 30% in hotels. The major reduction results from AI system prediction capabilities, which let machines forecast technical equipment failures before scheduling preventive maintenance. Hotels can safeguard continuous service delivery by handling potential problems before they reach critical levels to achieve better guest satisfaction and operational dependability. Maintenance optimization through AI means resources are used efficiently, and only required operations happen at the right times because systems operate with real-time data-based scheduling. Significant cost efficiencies arise from fewer avoidable maintenance tasks and emergency repairs. Energy efficiency stands as a main benefit of maintaining assets through AI-driven systems. The implementation of AI systems by hotels enables their properties to maintain energy conservation rates at 15%. The chapter provides comprehensive statistical evidence regarding performance differences that exist between conventional maintenance practices and AI-powered maintenance systems. Operationally extending the lifespan of assets remains insufficient by using traditional methods, thus requiring more immediate replacements at higher capital costs. A cost-effective operation emerges with higher efficiency through decreased labor expenses, increased productivity, and minimized operational disruptions. These new industry advancements have set new standards of quality excellence, enabling hotels to thrive in competitive markets by delivering outstanding experiences to their guests.

REFERENCES

Ali, F. and Amin, M. 2014. The influence of physical environment on emotions, customer satisfaction and behavioral intentions in Chinese resort hotel industry. Journal of Global Business Advancement. 7(3): 249–266. https://doi.org/10.1504/JGBA.2014.065071

Buhalis, D. and Amaranggana, A. 2015. Smart tourism destinations enhancing the tourism experience through personalisation of services. *In*: Tussyadiah, I. and Inversini A. (eds). Information and Communication Technologies in Tourism 2015. 377–389. Springer, Cham. https://doi.org/10.1007/978-3-319-14343-9_28

Buhalis, D. and Leung, R. 2018. Smart hospitality—Interconnectivity and interoperability towards an ecosystem. International Journal of Hospitality Management. 71: 41–50. https://doi.org/10.1016/j.ijhm.2017.11.011

Choi, S. and Kandampully, J. 2019. The impact of technology on the service encounter and customer service experience in hospitality. Journal of Hospitality and Tourism Management. 40: 17–19. https://doi.org/10.1016/j.jhtm.2019.04.010

Connolly, D.J. and Olsen, M.D. 2017. An environmental scanning framework for strategic planning in the hospitality industry. International Journal of Hospitality Management. 20(2): 111–125. https://doi.org/10.1016/S0278-4319(00)00033-0

Evans, N. 2021. Strategic Management for Hospitality and Tourism. Routledge. https://doi.org/10.4324/9781315548061

Gretzel, U., Sigala, M. and Koo, C. 2015. Smart tourism: Foundations and developments. Electronic Markets. 25(3): 179–188. https://doi.org/10.1007/s12525-015-0196-8

Guttentag, D. 2015. Airbnb: Disruptive innovation and the rise of an informal tourism accommodation sector. Current Issues in Tourism. 18(12): 1192–1217. https://doi.org/10.1080/13683500.2013.827159

Jones, P., Hillier, D. and Comfort, D. 2016. Sustainability in the hospitality industry. International Journal of Contemporary Hospitality Management. 28(1): 36–67. https://doi.org/10.1108/IJCHM-11-2014-0562

Kar, A.K. 2020. Modelling user-generated content to develop the 'digital service usage satisfaction model'. Information Systems Frontiers. 22(4): 1117–1136. https://doi.org/10.1007/s10796-019-09909-2

Kasavana, M.L. 2017. Managing Technology in the Hospitality Industry. American Hotel and Lodging Educational Institute. https://doi.org/10.1108/9781787142022

Krizaj, D. and Knežević Cvelbar, L. 2019. A review of innovation research in tourism. Tourism and Hospitality Management. 25(1): 1–21. https://doi.org/10.20867/thm.25.1.8

Lu, Y. and Cecil, A.K. 2016. Artificial intelligence and robotics in the hospitality industry. Journal of Hospitality Marketing and Management. 25(6): 651–655. https://doi.org/10.1080/19368623.2016.1177813

Mark Ross-Smith. 2016. How Hotels use Big Data to Generate New Revenues. https://www.analyticssteps.com/blogs/how-analytics-transforming-hospitality-industry

Rawal, Y.S., Pal, S., Bagchi, P. and Dani, R. 2020. Hygiene and safety: A review of the hotel industry in the era of COVID-19 pandemic. Bioscience Biotechnology Research Communications. 13(10): 79–83.

Rawal, Y.S., Soni, H., Dani, R. and Bagchi, P. 2022. A review on service delivery in tourism and hospitality industry through artificial intelligence. *In*: Proceedings of Third International Conference on Computing, Communications, and Cyber-Security: IC4S 2021. 427–436. Singapore: Springer Nature Singapore.

Sharma, S., Rawal, Y.S., Pal, S. and Dani, R. 2022. Fairness, Accountability, Sustainability, Transparency (FAST) of Artificial Intelligence in Terms of Hospitality Industry. *In*: Fong, S., Dey, N. and Joshi, A. (eds). ICT Analysis and Applications. Lecture Notes in Networks and Systems. Vol. 314. Springer, Singapore. https://doi.org/10.1007/978-981-16-5655-2_48

Sharma, S., Rawal, Y.S., Soni, H. and Batabyal, D. 2023, February. Technological impacts of AI on hospitality and tourism industry. *In*: Proceedings of International Conference on Data Science and Applications: ICDSA 2022. 1: 71–78. Singapore: Springer Nature Singapore.

Takuli, Y.S.R., Sharma, S., Takuli, S.S. 2025. AI enabled predictive energy management system: A sustainable future for the hospitality industry. Grenze International Journal of Engineering and Technology, January Issue. 4196–4202, Grenze ID: 01.GIJET.11.1.332_1

Ted, A. Fernandez. 2024. The Hackett Group announces strategic acquisition of leading Gen AI development firm Leeway Hertz. https://www.leewayhertz.com/ai-use-cases-in-hospitality/

Chapter 4

Revolutionizing Travel: The Impact of AI on the Tourism Industry

Nilanjan Ray*[1], Dillip Kumar Das[2], Debasish Batabyal[3] and Priyanka Kanjilal[4]

[1]Associate Professor, Faculty of Management Studies, JIS University, Kolkata

[2]Professor, Department of Tourism Management, The University of Burdwan

[3]Assistant Professor, Department of Travel and Tourism, Amity University, Kolkata

[4]Assistant Professor, PhD Scholar, Asansol Engineering College, JIS University, Kolkata

INTRODUCTION

The tourist industry is a typical case of using information technology (IT) to improve communication with potential clients through the use of location images. In today's market context, the tourist industry has integrated various elements by utilising information technology to minimise the involvement of middlemen in the selling of travel packages to buyers. The study aims to comprehend how corporate social responsibility (CSR) and artificial intelligence (AI) are reshaping the tourist experience and tourist industry. It is also feasible for a tourist to see a place

*For Correspondence: dr.nilanjanmgmt@gmail.com

DOI: 10.1201/9781003540083-4

without being physically present through augmented reality (AR). With the help of information technology and artificial intelligence, the tourism sector has grown significantly and has become a rapidly increasing industry in the contemporary Industry 4.0 era. Application of AI in the tourism sector is prompted by the digitalisation as well as expansion of the tourism sector. Indian tourism companies are applying AI for enhancing client communication, streamlining operations, as well as staying competitive. Worldwide distribution of services began with the advent of computerised reservation systems (CRSs) by airlines in the 1970s and other technological advances (Naqvi and Jia 2014). Artificial intelligence-based data analysis tools are assisting Indian tourism companies in gaining more information about their competitors, market trends, and customer behaviour. The data can be utilised by organisations to remain competitive in the market by taking strategic decisions regarding pricing, marketing strategies, and product development. Information technology and artificial intelligence have notably transformed the travel and tourism industry in India and globally (Technology in Tourism Management 2023, Sharma et al. 2023). Operations have changed significantly due to improved customer experience, better resource utilisation, and the ability to implement more effective revenue plans. AI is increasingly being used in the tourism industry in India due to the growing demand from customers and the digitalisation of platforms (Agarwal et al. 2024). As these technologies become more widespread, countries should expect more competition among travel destinations, improved operational efficiency, and a collaborative creation of the consumer experience to enhance their travel experience.

LITERATURE REVIEW

The emergence of information technology has greatly changed the way destination marketing is done. Travel agencies can now interact with clients from all parts of the world in ways that were not possible before. The way destinations market their facilities, events, and experiences in real time has been completely transformed by various channels such as social media, websites, and mobile applications. (Buhalis and Law 2008) argue that Web 2.0 technology is necessary for enabling user-generated content, which is important for destination marketing. Websites like TripAdvisor and Yelp have become important platforms where travellers can share their views and recommendations, affecting the travel decisions of other people. As per the research by (Knani et al. 2022) on the use of artificial neural networks, data mining, sentiment analysis, and future subjects like robotic services, AI is used to find new themes in research. According to Law et al. (2014) on the use of IT technologies in the tourist sector, marketers in the tourist sector use IT technologies that promote innovation in online booking, revenue management platforms, customer interactions, and decision-making processes to gain a competitive edge. According to (Wang et al. 2019) on the use of AI in the travel industry, the use of AI has benefited the travel industry. They consider the ways in which AI can benefit both businesses and travellers by providing travellers with the necessary information to make informed decisions and tailor their experiences.

Table 4.1 Application of AI in the travel and tourism sector

Applications of AI (Travel and Tourism)	Hallmarks
Chatbots	AI-powered chatbots are increasingly adopted in the travel industry by service providers to bolster customer support. While they may not handle every inquiry, these AI chatbots significantly reduce the workload of travel agents, allowing them to focus on more complex tasks
AI-enabled Smart Applications	AI tools and technology within the travel sector empowers travellers to create personalized travel itineraries. Using smart AI-powered applications, travel brands can offer guests customized recommendations that align with their preferences. These applications also integrate travel mapping services, aiding travellers in exploring and navigating their selected destinations effectively, easing travel sales and purchase analysis.
Voice-activated assistants	AI-driven voice assistants have made a notable impact on the travel and tourism industry. Through straightforward voice commands, tourists can effortlessly search for flights, hotels, and attractions, streamlining the planning process. These assistants also provide timely updates on travel information like weather conditions and traffic, enhancing overall convenience for users
Facial identification	Travel industry extensively uses face recognition technologies with AI to ensure safety and compliance with regulations. It saves a lot of time by accelerating processes at train stations, airports, hotels, restaurants, and large events. It increases travel efficiency and convenience for all parties by recognizing people by visual traits from previous visits saved in the database.
AI-enabled Virtual Tours	Virtual tour guides driven by AI will help travellers by providing them with accurate and helpful information. The experiences and viewpoints of tourists could be enhanced by these tour guides.
Personalized Recommendations	AI analyses user preferences and behaviour to suggest customized travel destinations, activities, and accommodations.
Image Recognition	Identifies landmarks, attractions, and user-generated content from images for enhanced travel planning and recommendations. It enriches content and improves travel experience.
Fraud Detection	Monitors transactions for unusual patterns and behaviours to prevent fraudulent activities. It increases security and reduces financial risks.
Travel Forecasting	Predicts travel trends and consumer behaviour to help businesses adapt and plan effectively. It supports strategic planning and market positioning.

While there are definitely some notable benefits to using robots in tourism, it is also important to keep in mind the challenges that come along with them. It is essential to find a good balance when integrating these robots so that the advantages truly outweigh the potential downsides, all while keeping visitors' happiness and overall experience a top priority. People working in the tourism industry need to understand how these robots function and how they can boost productivity and quality of service to create truly unforgettable experiences for visitors. Additionally, we have to ensure that we are protecting the rights and interests of both travellers and tourism workers, which means finding the right balance between automated interactions and human touch. It is all about balancing the benefits with any possible drawbacks.

Table 4.1 lists chatbots, smart assistants, and virtual tour systems as examples of AI applications in travel and tourism that improve operational safety, customization, and customer service. These technologies, which use automation and personalized recommendations, not only maximize service efficiency but also raise traveler happiness.

In the end, how robots are controlled and utilized will determine whether they benefit or negatively impact travel. Entrepreneurs, managers, leaders, scholars, and everyone in between have all recognised the necessity for innovative solutions, and one that has been mentioned as being full of potential is artificial intelligence. Moreover, the convergence of mobile and Wi-Fi technologies has significantly enhanced the travel experience by offering unparalleled convenience to tourists. These technologies (Table 4.2) enable seamless access to tour-related information through GSM, WAP, and GPRS networks, empowering travellers to efficiently book accommodations, transportation, and other essential travel services.

Table 4.2 Software used in the hotel industry

ASI Front Desk Freeware 6.0	With features including reservation administration, check-ins, billing, and reporting, ASI Front Desk Freeware 6.0 is a free hotel management programme designed for small hotels and guesthouses. Its goal is to increase efficiency through front-office automation without requiring a large capital outlay.
eZee Front Desk Hotel Software 5.2	It is a comprehensive hotel property management system that includes front-office operations, reservations, guest management, invoicing, and housekeeping. It may also include reporting, analytics, and integration with POS and online booking channels, all with the goal of increasing productivity and guest satisfaction for hotels of all sizes and types.
SIMSOFT Hotel pro 2006	All-in-one software for hotels that simplifies tasks like booking reservations, managing the front desk, coordinating housekeeping, handling bills and invoices, and generating reports
Room Master 2000 Hotel Reservation Software 12	Manage advanced reservations and deposits, email and printing confirmation for group blocks, and revenue management
EXSS Facility Manager 5.2	This software is intended primarily for small hotels, universities, fitness centres, SPAs, sports stadiums, gymnasiums, rent-a-cars, service areas, sanatoriums, pensions, rental facilities, and so on.

Source: (Ray et al. 2011) International Conference on Management and Service Science, Wuhan, China

Role of Chatbots

As artificial intelligence and natural language processing (NLP) technologies have advanced, chatbots have undergone significant evolution. Recently, travel and tourism organisations have successfully integrated chatbots with their established systems, such as booking engines and customer databases. This seamless integration has elevated chatbots into powerful tools adept at managing intricate travel-related inquiries and transactions with efficiency. The application of chatbots in the tourism sector has multiple uses. Figure 4.1 illustrates the application of chatbots in the tourism sector.

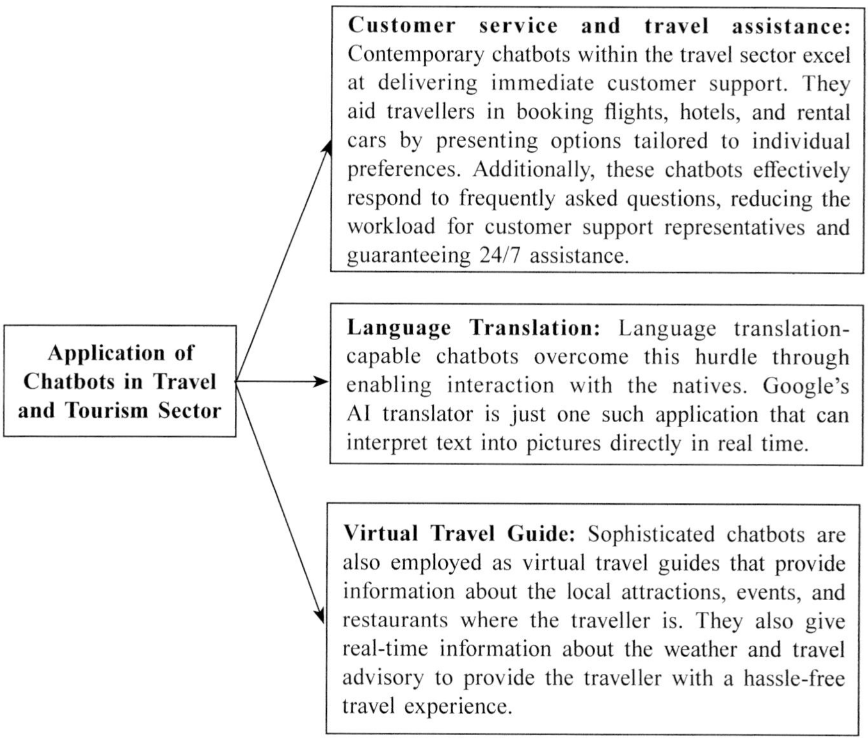

Figure 4.1 Role of chatbots in travel and tourism sector (*Source:* Author's compilation).

Role of AI-Enabled Virtual Assistants

AI-enabled software or apps helping travellers with their trip planning and experience are referred to as virtual assistants in tourism. The virtual assistants understand the queries of travellers and offer customised services and information based on machine learning and natural language processing. The major features of these AI-enabled Virtual Assistants are as follows (Table 4.3). Artificial intelligence (AI) algorithms based on machine learning are widely utilised to study the behaviour of the user, interests, and historical travel activities to suggest customised suggestions for destinations to visit, hotels, and activities. AI-enabled virtual assistants are utilised in customer service to aid in bookings and reply to questions on time. Apart from providing customised travel suggestions based on the user's preferences, these systems accomplish mundane tasks like switching bookings and responding to FAQs efficiently.

Table 4.3 Essential features of AI-enabled travel assistants

Feature	Description
Natural Language Processing (NLP)	Understands and processes user queries in natural language, enabling intuitive interactions.
Personalization	Customizes recommendations and travel plans based on user preferences, history, and behaviour.
Real-time Updates	Provides live information on flight statuses, weather conditions, and local events.
Multi-language Support	Offers assistance in multiple languages to cater to a diverse user base.
Integration with Booking Systems	Allows users to book flights, hotels, and other travel services directly through the assistant.
Itinerary Management	Organizes and manages travel itineraries, including schedules, reservations, and reminders.
Local Recommendations	Suggests local attractions, restaurants, and activities based on user interests and location.
Expense Tracking	Helps track and manage travel expenses and provide budget insights.
Contextual Awareness	Recognizes and adapts to the context of the conversation, such as time zones and user location.
Integration with Maps	Provides navigation and route planning using integrated maps and GPS

Source: Author's Compilation

CONCLUSION

The tourism industry has witnessed a radical transformation with the use of information technology. This has resulted in improved access and convenience of booking and utilisation of data analysis for providing personalised experiences to specific customers. The effectiveness and efficiency of business processes have also seen a marked improvement. The availability of smartphone applications, easy online booking facilities, and guidance through intelligent devices have also helped in improving trip planning. The airline and hotel industries are also feeling the effects of a new trend called hyper-personalisation. Its objective is to reverse the decline of customer loyalty and retention through the use of artificial intelligence. With the convergence of AI and IT, the tourism industry has seen a tremendous revolution in consumer interaction, service, and decision-making. Convergence has changed destination marketing strategies, operational modes, and consumer interactions through the provision of personalised travel itineraries, augmented reality experiences, and simplified booking processes. Future research might examine how AI and IT may assist in sustainable tourism and environmentally friendly travel, particularly in sustainable travel and environmental conservation.

The research agenda is intended to cover the application of Web GIS and Geospatial Mashups as management and analytical tools for tourism management and marketing.

REFERENCES

Abreu, A., Costa, E. and Borges, I. 2020a. The Impact of Artificial Intelligence on the Tourism Industry: A Systematic Review. Advances in Tourism, Technology and Systems: Selected Papers from ICOTTS20. 1(208): 458.

Abreu, L.H.P., Pimentel, M.C.L., Borges, W.F.A., Costa, T.H.C., Naeem, M., Iqbal, J., et al. 2020b. Plasma nitriding of AISI M2 steel: Performance evaluation in forming tools. Surface Engineering. 36(5): 508–515.

Agarwal, L., Bolia, N.B. and Gupta, R. 2024. Route guidance map for emergency evacuation using intelligent transportation analytics. International Journal of Advanced Research in Engineering and Technology. 15(4): 112–124.

Batabyal, D., Rawal, Y.S., Chatterjee, P. and Goswami, S. 2023. Temple service quality model for future pilgrimages in Indian sub-continent through user-generated contents (UGC) in TripAdvisor. International Journal of Tourism Anthropology. 9(2): 113–135.

Buhalis, D. and Law, R. 2008. Progress in information technology and tourism management: 20 years on and 10 years after the internet—The state of eTourism research. Tourism Management. 29(4): 609–623.

Bulchand-Gidumal, J. and Melián-González, S. 2020. Why are ratings so high in the sharing economy? Evidence based on guest perspectives. Current Issues in Tourism. 23(10): 1248–1260. https://doi.org/10.1080/13683500.2019.1602597

Bulchand-Gidumal, J., William Secin, E., O'Connor, P. and Buhalis, D. 2023. Artificial intelligence's impact on hospitality and tourism marketing: Exploring key themes and addressing challenges. Current Issues in Tourism. 0(0): 1–18. https://doi.org/10.1080/13683500.2023.2229480

Chaudhuri, S. and Ray, N. 2015. Application of web-based Geographical Information System (GIS) in tourism development. *In*: Encyclopedia of Information Science and Technology, 3rd Ed. 7026–7036. IGI Global.

Chaudhuri, S. and Ray, N. (eds). 2018. GIS Applications in the Tourism and Hospitality Industry. IGI Global.

Cho, Y.H. and Fesenmaier, D.R. 2001. A new paradigm for tourism and electronic commerce: Experience marketing using the virtual tour. Business, Computer Science.

Corte, V.D., Gaudio, G.D., Sepe, F. and Sciarelli, F. 2019. Sustainable tourism in the open innovation realm: A bibliometric analysis. Sustainability. 11: 6114.

De Carvalho, J.V., Rocha, Á., Liberato, P. and Peña, A. (eds). 2021. Advances in Tourism, Technology and Systems: Selected Papers from ICOTTS20. 1(208). Springer Singapore. https://doi.org/10.1007/978-981-33-4256-9

García-Madurga, M.-Á. and Grilló-Méndez, A.-J. 2023. Artificial intelligence in the tourism industry: An overview of reviews. Administrative Sciences. 13(8): Article 8. https://doi.org/10.3390/admsci13080172

Jakkilinki, R., Georgievski, M. and Sharda, N. 2007. Connecting destinations with an ontology-based e-Tourism planner. *In*: Sigala, M., Mich, L. and Murphy, J. (eds). Information and Communication Technologies in Tourism 2007. 21–32: Springer-Verlag Wien.

Jovanović, V. 2016. The application of GIS and its components in tourism. Yugoslav Journal of Operations Research. 18(2).

Knani, M., Echchakoui, S. and Ladhari, R. 2022. Artificial intelligence in tourism and hospitality: Bibliometric analysis and research agenda. International Journal of Hospitality Management. 107: 103317.

Law, R., Buhalis, D. and Cobanoglu, C. 2014. Progress on information and communication technologies in hospitality and tourism. International Journal of Contemporary Hospitality Management. 26(5): 727–750.

Li, C.Y., Le, J.B., Wang, Y.H., Chen, S., Yang, Z.L., Li, J.F., et al. 2019. *In situ* probing electrified interfacial water structures at atomically flat surfaces. Nature Materials. 18(7): 697–701.

Naqvi, M.A. and Jia, H. 2014. Computer Reservation System, tourism. Encyclopaedia of Tourism. Springer, Cham. https://doi.org/10.1007/978-3-319-01669-6_510-1. Retrieved on October, 11, 2021.

Palomoa, J., Figueroa-Domecqa, C., FlechaBarrioa, M.D. and Segovia-Pérez, M. 2017. The use of new data analysis techniques in tourism: A bibliometric analysis in data mining, big data and structural equations models. e-Review of Tourism Research.

Rawal, Y.S., Başer, M.Y., Batabyal, D. and Durmaz, Y. 2024. The mediating role of purchasing duration in the effect of social media dependency on consumer purchasing behavior: Comparison of India and Turkey. Pacific Business Review (International). 17(5): 18–33.

Ray, N., Das, D.K., Ghosh, S. and Banerjee, A. 2011. A study on application of information technology on tourism development. IEEE Proceeding Catalog No: CFP1141H-ART ISBN978-1-4244-6579- China.

Ray, N., Dey, P., Chakraborty, M. and Bandyopadhyay, A. 2024. A study on AI application in travel and tourism industry. *In*: Das, G. (ed.). Redefining Commerce and Management: New Paradigms for the Digital Age. 1: 222.

Samala, N., Katkam, B.S., Bellamkonda, R.S. and Rodriguez, R.V. 2020. Impact of AI and robotics in the tourism sector: A critical insight. Journal of Tourism Futures. 8(1): 73–87.

Seguí-Amortegui, L., Clemente-Almendros, J.A., Medina, R. and Grueso Gala, M. 2019. Sustainability and competitiveness in the tourism industry and tourist destinations: A bibliometric study. Sustainability. 11(22): 6351. https://doi.org/10.3390/su11226351

Serrano, L., Sianes, A. and Ariza-Montes, A. 2019. Using bibliometric methods to shed light on the concept of sustainable tourism. Sustainability. 11(24): 6964.

Sharma, S., Rawal, Y.S., Soni, H. and Batabyal, D. 2023, February. Technological impacts of AI on hospitality and tourism industry. *In*: Proceedings of International Conference on Data Science and Applications: ICDSA 2022. 1: 71–78. Singapore: Springer Nature Singapore.

Singh, S. and Bashar, A. 2021. A bibliometric review on the development in e-Tourism research. International Hospitality Review. 37(1): 71–93.

Wang, J., Li, Y., Hu, G. and Yang, M. 2019. Lightweight research in engineering: A review. Applied Sciences. 9(24): 5322. https://doi.org/10.3390/app9245322

Xiang, Z., Du, Q. and Ma, Y. 2017. ICT-enabled seamless travel: A conceptual framework and its applications. Journal of Hospitality and Tourism Technology. 8(1): 63–75.

Chapter 5

Safety and Security in Tourism through AI Applications

Anil Chandhok*[1], Padam Bhushan[2] and Ananya Jain[3]

[1]Professor, University School of Business, Chandigarh University, Mohali, India

[2]Associate Professor, University School of Business, Chandigarh University, Mohali, India

[3]BBA 3rd Semester, University School of Business, Chandigarh University, Mohali, India

INTRODUCTION

Safety and security are the key issues and are the basic pillars of the tourism sector, essential for fostering a positive travel experience and ensuring sustainable growth. The tourism industry, which to a great extent contributes to the global economy, depends heavily on the perception and reality of safety to attract and retain tourists and keep them delighted so that they can share their good experiences with their relatives and other close reference groups. Any threats to safety, whether through terrorism, natural disasters, or cyberattacks, can severely damage a destination's reputation, leading to economic downturns and to the loss of livelihoods for those who are totally dependent on tourism-related services (Zou and Zheng 2020).

Moreover, it is an ethical responsibility of tourism business players to protect their customers from all sorts of risks and harm and make safety a top priority for them. At the global level, where world travel is continuously increasing, the

*For Correspondence: anil.e4863@cumail.in, padam.e11878@cumail.in, 22bba10562@cuchd.in

DOI: 10.1201/9781003540083-5

need for robust safety and security measures has never been more critical than before as the population and number of tourists have increased extensively. These measures are not only going to protect the physical well-being of the tourists and travellers but also contribute towards the psychological comfort and confidence needed to enjoy travel. One of the most common AI applications in tourism safety is predictive analytics, which helps in forecasting potential risks, such as natural disasters, disease outbreaks, or security threats, by analysing vast datasets (Tussyadiah 2020, Ghosh and Batabyal 2023).

AI-driven emergency response systems optimise crisis management by analysing real-time data from multiple sources to coordinate rapid responses in emergencies, such as natural disasters or terrorist threats (Liu et al. 2022). AI-based behavioural analysis tools are also being used to detect suspicious activities in crowded areas, thereby enhancing public security (Chen et al. 2021). AI-powered cybersecurity systems also help protect sensitive tourist data, including payment information and personal identification, from potential cyber threats and fraud (Buhalis and Law 2008).

Ensuring safety and security is therefore now becoming indispensable for the sustainability and prosperity of the tourism sector, safeguarding both travellers and tourist destinations.

ARTIFICIAL INTELLIGENCE APPLICATION FOR THE SAFETY AND SECURITY WITH REFERENCE TO THE TOURISM SECTOR

Overview of AI in Tourism Safety and Security

AI tools practised in the tourism sector have primarily focused on improving the overall operational efficiency of tourism-related services, enhancing the client experience, and ensuring a high degree of safety and security for their lives. The usage of AI applications ranges from predictive analytics and surveillance to emergency response systems and cybersecurity. Early studies undertaken, such as those by (Buhalis and Leung 2018), emphasise the potential and prospects of AI to predict and mitigate risks in real-time, making it an invaluable asset for the tourism sector.

AI in Predictive Analytics and Risk Management

Predictive analytics, powered by AI tools, is one of the most prominent applications in extending safety to all stakeholders. According to (Tussyadiah and Park 2018), AI algorithms can analyse massive amounts of big data generated from various sources, including social media, booking platforms, and weather reports, to predict likely threats such as natural disasters like floods, earthquakes, cloud bursts, terrorist activities, or health epidemics. These predictive analytics allow tour and travel operators and governments to frame proactive measures for protecting tourists and minimising the chances of risks to life (Matyas and Pohl 2018).

AI-driven tools and applications are now being extensively practised to predict the spread of diseases, as noticed during COVID-19 pandemic phase. A study by (Yang et al. 2021, Ghosh and Batabyal 2023) demonstrated how AI tools could analyse the travel patterns and health data to forecast the outbreak of hotspots, enabling authorities to plan and implement the required health and safety measures to face the situation.

AI in Surveillance and Monitoring

AI-integrated surveillance systems are frequently practised and have become a vital instrument in assuring and extending security at tourism destinations. According to research undertaken by Korstanje and Skoll (2019), AI-supported high-tech cameras and sensors can monitor large crowds, like at Maha Kumbh at Prayagraj, detect suspicious activities, and alert security personnel to be proactive in meeting the demands of the situation. These systems are particularly useful for carrying out surveillance over crowded places, including through drones at airports and popular tourist attractions, where manual monitoring would not be adequate.

Facial reading is another AI application extensively practiced within the tourism sector for the purpose of identifying, verifying and assessing control. As pointed by (Bowen and Whalen 2017), this face recognition technology not only adds to the safety and security of the tourists and other stakeholders but also adds to the efficiency of check-in processes, thereby contributing to a safer and smoother travel experiences for tourists in general.

AI in Emergency Response

The usage of AI applications in emergency response is also gaining significant recognition. Intelligent systems can analyse the vast volume of data generated from multiple sources, such as satellite images, social media platforms, and sensor networks, to provide real-time information during crisis periods. A study undertaken by (Zeng et al. 2019) highlighted the use of AI in disaster management, focusing on how AI systems can predict natural disasters, optimise evacuation routes, and coordinate rescue operations more effectively to mitigate risks and losses (Zeng et al. 2019; Noble 2018).

From the context of terrorism, AI applications are being practised to detect and prevent potential attacks. For instance, advanced AI algorithms can analyse communication patterns, financial transactions, and social media activity to identify and track potential threats before they actually happen or occur (Klein and Williams 2020).

AI in Cybersecurity

With the rising dependence on digital platforms in tourism, cybersecurity has become a major issue for today and for the future. AI technologies are increasingly being used to protect sensitive and confidential data from cyber threats. According to a review by (Morakanyane and Grace 2020), AI can detect and identify potential security breaches and respond to cyber-attacks. These capabilities are critical for protecting not only sensitive personal data but also the financial integrity of tourism stakeholders.

As dependency on online platforms has increased over the years, AI-driven fraud detection systems are considered essential in preventing financial fraud through online bookings and transactions, safeguarding both parties, i.e., clients and businesses in the tourism sector (Batabyal and Das 2013; Davenport and Ronanki 2018).

APPLICATIONS OF AI TOOLS IN THE TOURISM SECTOR FOR SAFETY AND SECURITY PURPOSES

The application of Artificial Intelligence (AI) in the field of tourism is notably improving safety and security concerns, ensuring that tourists can enjoy their experiences with greater peace of mind. AI applications in this field are multifaceted, addressing everything from fraud prevention to safety monitoring. Here are some dimensions where AI is being applied to extend safety and security to different stakeholders involved in the tourism industry:

Fraud Detection and Prevention

- *AI in Payment Security:* AI algorithms analyse the transaction patterns to detect unusual activity, helping to prevent fraud through bookings on online platforms, payments, and reservations. By identifying the anomalies, AI can detect or fish out the suspicious transactions, ensuring that fraudulent activities are minimized (Neuhofer et al. 2015).
- *Identity Verification:* AI-integrated devices/systems use facial recognition and biometric data to authenticate travellers' identities, reducing the risk of theft and fraud resulting from the use of forged travel documents.

Enhanced Airport Security

- *Facial Recognition Technology:* In India, airport authorities are practising AI-powered facial recognition systems to streamline passenger check-ins, boarding processes, and customs checks. This technology not only adds to promptness but also increases security.
- *Baggage Screening:* AI is also being practised for the automated scanning of luggage, identifying prohibited items more efficiently and reducing the risk of harmful and inflammable materials being carried on flights. Machine learning models help in improving accuracy in these matters (Nguyen and Tran 2021; Zou and Zheng 2020).

Predictive Analytics for Risk Management

- *Travel Risk Assessments:* AI uses predictive analytics to assess potential risks associated with specific travel destinations, such as political instability, natural disasters, or health outbreaks. Such information can be used by travel agencies, tour operators, and individual travellers to make rational and sound decisions about where and when to travel safely (Singh and Kumar 2022).
- *Crowd Control:* AI-integrated systems monitor and predict crowd movements at popular tourist sites like airports and other public spaces such as railway stations or religious events like Maha Kumbh at Prayagraj (Safe Tourism Project 2023). These AI-driven technologies help manage crowds to prevent overcrowding, which can be a security risk, and ensure that necessary emergency services can be made available efficiently if required.

Health and Safety Monitoring

- *AI in Pandemic Response:* During the COVID-19 pandemic, AI tools were instrumental in monitoring and controlling the spread of coronavirus in society (UNWTO Report 2023). AI systems were extensively used to track health data, enforce quarantine protocols, and ensure compliance with health regulations at airports and hotels.
- *Wearable Health Devices:* AI-driven wearable devices can monitor the health conditions of tourists and travellers and alert the concerned authorities in case of any potential health issues that may arise. Such devices can monitor vital signs, track symptoms, and even predict potential illness outbreaks in specific locations.

Intelligent Surveillance Systems

- *Smart Cameras and Video Analytics:* AI-equipped high-tech surveillance cameras at airports, hotels, and tourist attractions are being used for advanced video analytics to detect suspicious behaviour of individuals or unauthorised access (India Tourism Statistics at a Glance 2023). These systems can alert security personnel to potential threats before they actually take place.
- *Behavioural Analysis:* AI-supported systems are also being used to analyse the behaviour of individuals and groups in public places to detect unusual activities that might indicate security risks, such as loitering, unattended luggage, or erratic movements.

Cybersecurity

- *Protection of Sensitive Data:* The tourism industry, including peripheral services, deals with vital personal and financial data, making it a target for cyber-attacks. AI-based cybersecurity systems monitor networks for potential threats and protect against data breaches, ensuring the security of information for all stakeholders.

- *AI in Securing Online Travel Platforms:* AI is used to secure online travel platforms by identifying and mitigating threats such as phishing attacks, fake reviews, and unauthorised access to user accounts.

Emergency Response and Disaster Management

- *AI Applications in Crisis Prediction:* AI algorithms are also used for analysing environmental and social data to forecast potential crises such as natural disasters, terrorist attacks, or civil unrest. This makes it possible to make provisions for proactive actions, such as issuing travel advisories or initiating evacuation plans so that the impact of the crisis can be minimised if it cannot be prevented.
- *Real-Time Communication:* AI-powered chatbots and virtual assistants can provide tourists and travellers with required information during emergencies, including evacuation routes, emergency contacts, and safe zones (Bulchand-Gidumal and Secin 2023). These tools ensure that tourists receive accurate and timely information when it is most needed.

AI Contribution in Travel Insurance

- *Risk Assessment for Insurance:* AI applications also help travel insurance companies assess the risk profiles of tourists more accurately, offering customised policies based on their travel history, health status, and destination risks (Jin and Zhang 2021). This ensures that tourists and travellers have adequate coverage for potential safety and security concerns during their trip.
- *Automated Claims Processing:* In case of any incident, AI can speed up the claims settling process by automatically verifying the incidents, processing claims, and issuing payouts, thereby reducing stress on tourists during difficult situations.

Border Security

- *AI-Powered Border Control:* AI systems are also being practised in border control processes to enhance safety and security efficiently. Automated border gates equipped with facial recognition and biometric scanning ensure that only authorised individuals are allowed entry, reducing the risk of illegal entry.
- *Advanced Screening Techniques:* AI-driven devices assist in screening travel documents, detecting forged or altered passports, visas, and other identification papers with high accuracy and efficiency (Gopal and Varma 2019).

Proactive Communication Systems

- *Personalized Alerts:* AI tools can provide tourists with personalised safety alerts based on their current location, upcoming travel plans, and the specific risks associated with their destination. These alerts can include weather warnings, health advisories, and security updates.
- *Language Translation:* In emergencies, AI-driven translation apps can help tourists communicate with local authorities and service providers, ensuring they receive the help they need even if they do not speak the local language.

Translation Apps

Such apps are used for translating one language into another to understand the correct meaning and avoid situations that may cause harm, delays, or confusion. Such apps assist tourists in navigating unfamiliar environment and facilitate conversation in multiple languages.

Google Translate: Offers real-time translations.

iTranslate Voice: Facilitates conversations in multiple languages.

Predictive and Preventative Maintenance

AI applications are also intensively used for monitoring the structural health of attractions, transportation systems, and hotels to prevent mishappenings and accidents (Chien and Chen 2022). AI-based drones are also being used for regular inspections of bridges, trails, or monuments. Such tools help to avoid disasters that could otherwise take the lives of innocent tourists and even the lives of common people.

Enhanced Identity Verification and Fraud Prevention

Practicing advanced facial recognition for seamless and secure access at airports, hotels, and tourist attractions is common to avoid the entry of wrong people with intentions to harm others or their property.

CHALLENGES AND ETHICAL CONSIDERATIONS

- *Privacy Concerns:* The usage of AI-enabled tools for extending safety and security, especially involving the use of biometric data and surveillance, raises significant privacy concerns. Balancing security needs with the protection of personal privacy is a critical issue to be addressed.
- *Bias in AI Algorithms:* There is a risk that AI algorithms used in security applications may be biased, leading to discriminatory practices. Fairness and transparency in AI-driven security systems are therefore essential.

Table 5.1 Summary of the various AI tools and their usages for the safety and security in the tourism sector

AI Tool/Technology	Application	Description	Example
Predictive Analytics	Risk Prediction and Management	Analysing data to forecast potential threats such as natural disasters or health outbreaks.	Predictive models for disease outbreaks.
Facial Recognition	Identity Verification and Surveillance	Identifies and verifies individuals, enhances security at airports and hotels, and streamlines check-ins.	Airports are using facial recognition for border control.
Chatbots and Virtual Assistants	Customer Service and Support	Provides 24/7 assistance, answers queries, and handles bookings, enhancing safety and convenience.	AI chatbots on hotel websites.
Cybersecurity Systems	Data Protection and Fraud Detection	Detects and responds to cyber threats, protects sensitive data, and prevents online fraud.	AI systems for detecting fraudulent transactions.
AI-Driven Risk Assessment Tools	Threat Detection and Assessment	Evaluates potential risks and vulnerabilities in travel plans, helping to mitigate threats.	Risk assessment tools for travel advisories.
Emergency Response Systems	Crisis Management and Coordination	Analyses data from various sources to coordinate emergency responses and optimize evacuation routes.	AI in disaster response for optimizing rescue operations.
Behavioural Analysis Tools	Suspicious Behaviour Detection	Analyses patterns in behaviour to identify potential security threats or unusual activities.	AI systems are monitoring crowd behaviour in public places.
Booking Optimization	Dynamic pricing algorithms and predictive analytics	Cost optimization for travellers and operators	Hopper, Skyscanner

Source: Self-made Table for a brief summary

- *Dependence on AI:* Over-dependence on AI systems for safety and security could lead to challenges in situations where these systems fail or are compromised. It is important that efforts be made to maintain a balance between AI driven devices and the human element.

FUTURE OF AI APPLICATIONS FOR SAFETY IN TOURISM SECTOR

The prospects of AI-driven tools and their usage for the safety and security of travellers and tourists hold immense promise, driven by ongoing advancements in technology and increasing demands for enhanced protection and efficiency. AI tools will use advanced algorithms to analyse historical and real-time data to forecast potential threats such as natural disasters, health crises, or security incidents. This will enable service providers to be more proactive in protecting tourists before incidents actually take place (Cao and Liu 2023). AI-powered surveillance systems will offer improved accuracy and efficiency in monitoring public crowded spaces like the Maha Kumbh at Prayagraj. These systems will use facial recognition, behaviour analysis, and anomaly detection to identify and respond to security threats quickly.

AI will integrate with IoT devices to develop an intelligent infrastructure in tourist destinations. This includes smart hotels with AI-driven safety features, connected transportation systems, and intelligent public safety systems. AI-driven language translation apps will make communication between tourists and local services feasible, reducing the chances of misunderstandings and improving safety in unknown and foreign environments. It can be said that the use of AI for the safety and security of tourists will surely increase in the because of its high degree of contribution.

The tourism industry's adoption of AI applications will not only elevate convenience and personalization but also address the most critical challenges like sustainability and safety, paving the way for a smarter, safer, and more inclusive future for all stakeholders in the tourism sector. Table 5.1 provides a summary of various AI tools and their usage for the safety and security in the tourism sector.

REFERENCES

Artificial Intelligence for India: Opportunities in Public Safety and Security, published by National Strategy for AI, NASSCOM, website: https://nasscom.in

Batabyal, D. and Das, D.K. 2013. Management of travel booking in the light of destination marketing: Evidence from an Alpine Indian State-Sikkim. Indian Journal of Marketing. 43(11): 36–45.

Bowen, J. and Whalen, E.A. 2017. Trends that are changing travel and tourism. Worldwide Hospitality and Tourism Themes. 9(6): 592–602.

Buhalis, D. and Leung, R. 2018. Smart hospitality—Interconnectivity and interoperability towards an ecosystem. International Journal of Hospitality Management. 71: 41–50.

Buhalis, D. and Law, R. 2008. Progress in information technology and tourism management: 20 years on and 10 years after the Internet—The state of eTourism research. Tourism Management. 29(4): 609–623.

Cao, S. and Liu, Y. 2023. Smart tourism and AI: Enhancing safety through technological innovation. Journal of Smart Tourism. 5(1): 34–47.

Chien, C.C. and Chen, Y.Y. 2022. Artificial intelligence in tourism: The application of chatbots for customer service and security. Tourism Management Perspectives. 42: 101–109.

Chen, Y., Zhang, X. and Wang, L. 2021. AI-powered surveillance for public security in tourism hotspots. Journal of Travel Security. 15(3): 45–60.

Davenport, T.H. and Ronanki, R. 2018. Artificial Intelligence for the real world. Harvard Business Review. 96(1): 108–116.

Ghosh, B.K. and Batabyal, D. 2023. Risk perception and travel behavior of urban residents in a densely populated region of East India Amidst (COVID)-19 Pandemic. Tourism Culture and Communication. 23(2–3): 193–205.

Gopal, R. and Varma, A. 2019. Technology-driven safety in Indian tourism: The case for AI integration. Indian Journal of Tourism Management. 15(2): 45–57.

India Tourism Statistics at a Glance. 2023. Published by: Ministry of Tourism, Government of India, website: https://tourism.gov.in

Jacques Bulchand-Gidumal and Eduardo William Secin 2023. Artificial Intelligence's Impact on Hospitality and Tourism Marketing. Journal of Travel and Tourism Marketing. 40(1): 2023.

Jin, X. and Zhang, Y. 2021. AI-driven risk management in tourism: A case study of predictive analytics and threat detection. Journal of Risk Research. 24(6): 789–802.

Klein, N. and Williams, T. 2020. Counter-terrorism and tourism: Implications of the AI revolution. Journal of Strategic Security. 13(4): 67–82.

Korstanje, M.E. and Skoll, G.R. 2019. Terrorism, tourism, and worker unions: The clash between capitalism and radical Islamism. Current Issues in Tourism. 22(11): 1360–1366.

Liu, M., Wang, R. and Zhao, J. 2022. Artificial intelligence in crisis management: Enhancing disaster response in tourism. Disaster Resilience Journal. 9(4): 112–127.

Morakanyane, R. and Grace, A. 2020. Cybersecurity in tourism: The role of artificial intelligence. Tourism Review. 75(2): 329–343.

Matyas, D. and Pohl, J. 2018. Ethical implications of AI in tourism security. Journal of Tourism Futures. 4(1): 45–54.

Noble, S.U. 2018. Algorithms of oppression: How search engines reinforce racism. New York University Press.

Neuhofer, B., Buhalis, D. and Ladkin, A. 2015. Smart technologies for personalized experiences: A case study in the hospitality domain. Electronic Markets. 25: 243–254.

Quoc-Loc Nguyen and Phi-Phung Tran 2021. The Role of AI in Shaping Future Tourism and Hospitality Trends. International Journal of Research in Engineering, Science and Management. 4(4): 2021.

Safe Tourism Project: Leveraging Technology for Security, Published by: NITI Aayog in 2023, website: https://www.niti.gov.in

Singh, A. and Kumar, R. 2022. The role of technology in reviving tourism post-pandemic: A focus on AI in India. Journal of Travel Research India. 10(1): 65–78.

Tussyadiah, I. 2020. Predictive analytics for risk assessment in tourism: A machine learning approach. Annals of Tourism Research. 83: 102962.

Tussyadiah, I. and Park, S. 2018. Artificial intelligence in travel and tourism. Tourism Management Perspectives. 27: 123–130.

UNWTO Report. 2023. Harnessing AI for Sustainable Tourism Development in Asia.

Yang, Y., Zhang, H. and Chen, X. 2021. COVID-19, big data, and AI in travel and tourism. Sustainability. 13(4): 1–14.

Zeng, Z., Liu, Z. and Shi, C. 2019. The application of artificial intelligence in emergency management of sudden disasters. Natural Hazards. 99(3): 1265–1279.

Zou, Y. and Zheng, X. 2020. Safety and Security in Tourism. *In*: Jafari, J. and Xiao, H. (eds). Encyclopedia of Tourism. 1–2. Springer International Publishing. https://doi.org/10.1007/978-3-319-01669-6_166-2

Chapter **6**

AI-Enhanced Tourism Experiences: Mediating Effects of Trust, Satisfaction, and Loyalty in Personalised Service Delivery

Ashish Adholiya*[1] and Shilpa Adholiya[2]
[1]Associate Professor, PAHER University, Udaipur (Rajasthan)
[2]Data Analyst and Researcher, Paas-Pro Educational Institute, Udaipur (Rajasthan)

INTRODUCTION

The rapid evolution of artificial intelligence (AI) has significantly impacted the global hospitality and tourism industries, transforming how businesses operate and how customers experience services. AI-driven innovations such as machine learning, natural language processing, and robotics have reshaped the landscape of hospitality by enhancing customer engagement, optimizing service delivery, and streamlining operational efficiency. From personalised recommendations and automated customer service to predictive analytics and dynamic pricing, AI has become an integral component of modern tourism management (Topsakal 2025, Patrichi 2024, Shahrizoda and Nargiza 2024, Gretzel and Koo 2021). As travellers increasingly seek seamless and customized experiences, the adoption of AI-driven

*For Correspondence: ashishadholiya@pacific-university.ac.in, asia_1982@rediffmail.com

DOI: 10.1201/9781003540083-6

technologies has allowed businesses to cater to their evolving needs, ultimately fostering higher levels of satisfaction and loyalty (Ivanov and Webster 2019).

Beyond efficiency, AI-driven personalisation has emerged as a game-changer in the tourism sector, offering tailored experiences that cater to individual preferences and expectations. Through data analytics and machine learning, AI enables businesses to understand consumer behaviour and anticipate their needs, delivering services that are both relevant and engaging (Tussyadiah 2020). This personalisation not only enhances the customer experience but also strengthens the relationship between businesses and travellers, leading to long-term customer retention and brand loyalty (Pencarelli 2020). However, the increasing reliance on AI in hospitality services also raises concerns about data privacy, ethical implications, and the potential loss of human interaction in service encounters.

This paper explores the role of AI in hospitality and tourism, with a specific focus on AI-driven personalisation and its impact on customer satisfaction and loyalty. It also examines the ethical considerations surrounding AI adoption and discusses future prospects for AI in the industry. By analysing these aspects, this study aims to provide a comprehensive understanding of how AI is shaping the future of tourism and hospitality while balancing technological innovation with human-centric service experiences.

Artificial Intelligence in Hospitality and Tourism

Artificial intelligence (AI) has emerged as a transformative force in the hospitality and tourism industries, reshaping service delivery, customer engagement, and operational efficiency (Avula and Sithole 2024). AI technologies, including machine learning, natural language processing, and robotics, have revolutionized how businesses interact with travellers. These advancements allow hotels, airlines, and travel agencies to provide seamless and highly personalised services (Gursoy 2025, Moghaddasnian 2024). AI-powered chatbots, virtual assistants, and automated booking systems have streamlined processes, reducing human error and enhancing customer convenience (Gretzel and Koo 2021). Additionally, AI-driven data analytics enable businesses to predict consumer behaviour, optimize pricing strategies, and improve service offerings, leading to higher guest satisfaction and loyalty (Ivanov and Webster 2019). As AI continues to advance, its role in the tourism sector is expanding beyond operational efficiency to shaping unique and memorable customer experiences.

AI-Driven Personalisation in Tourism

Personalisation has become a cornerstone of modern tourism, with AI playing a crucial role in tailoring experiences to individual preferences. AI algorithms analyse customer data, including browsing history, past purchases, and travel behaviours, to deliver customised recommendations. AI-driven platforms, such as Expedia and Airbnb, utilise machine learning to suggest personalised travel itineraries, accommodations, and activities (Koo et al. 2025, Semwal et al. 2024,

Tussyadiah 2020). AI-powered smart assistants in hotels can adjust room settings, recommend dining options, and offer local attractions based on guest preferences, enhancing overall satisfaction (Li et al. 2023). This level of personalisation strengthens the connection between travellers and service providers, making the experience feel more intuitive and customer-centric. By leveraging AI's predictive capabilities, businesses can anticipate needs even before they arise, further deepening customer engagement and fostering long-term loyalty (Pencarelli 2020).

Impact of AI-Enabled Service Attributes on Customer Satisfaction and Loyalty

The integration of AI in hospitality not only enhances service attributes such as responsiveness, reliability, and customisation but also plays a pivotal role in defining customer satisfaction and loyalty. AI-powered chatbots and virtual concierges provide 24/7 support, addressing customer queries in real time and reducing wait times (Chotisarn and Phuthong 2025, Singh and Singh 2024, Huang and Rust 2021). These innovations contribute to a frictionless service experience, where guests feel valued and attended to at every stage of their journey. Furthermore, AI-driven sentiment analysis helps businesses assess customer feedback, allowing them to refine service offerings and address potential concerns proactively (Kabadayi et al. 2019). This continuous feedback loop ensures that hospitality services remain dynamic and responsive to evolving consumer expectations.

While AI enhances efficiency, it is important to strike a balance between automation and human interaction. Guests still seek the warmth and emotional intelligence of human service providers, particularly in high-end or luxury hospitality settings (Amtul et al. 2025, Rane 2023, Bowen and Morosan 2018). Therefore, AI should be viewed as an enabler of human service rather than a replacement, with the goal of augmenting the guest experience rather than depersonalising it.

Ethical Considerations and Future Prospects of AI in Hospitality

Despite its benefits, AI adoption in hospitality raises ethical concerns, particularly regarding data privacy and security. The extensive use of AI-driven personalisation relies on collecting and processing large volumes of personal data, necessitating stringent data protection measures (Buhalis and Sinarta 2019). Transparency in data usage and obtaining customer consent are essential to maintaining trust between businesses and travellers. Moreover, concerns about job displacement due to AI automation highlight the need for a balanced approach that integrates AI without compromising human employment (Tenakwah and Watson 2025, George 2024, Ernst et al. 2019).

As AI technology continues to evolve, the future of hospitality and tourism will likely see even more sophisticated AI applications, such as emotion recognition systems, AI-driven virtual reality experiences, and sustainable tourism solutions powered by AI analytics (Rane et al. 2023, Mishra et al. 2024, Gretzel 2021). These advancements will enable businesses to craft deeply immersive and highly responsive travel experiences, reinforcing customer loyalty and satisfaction. Organisations that strategically implement AI while addressing ethical concerns will gain a competitive advantage, ensuring long-term success in the evolving tourism landscape.

RESEARCH QUESTIONS

Artificial Intelligence (AI) has revolutionised the hospitality and tourism industry, enhancing personalisation, service efficiency, and customer satisfaction (Gretzel and Koo 2021). AI-driven personalisation enables businesses to tailor services to individual preferences, improving guest experiences and fostering long-term loyalty (Pencarelli 2020). Moreover, AI-powered chatbots and virtual assistants streamline service interactions, reducing wait times and enhancing customer convenience (Huang and Rust 2021). Despite its benefits, the adoption of AI in hospitality presents challenges, including trust issues, ethical considerations, and the balance between automation and human interaction (Buhalis and Sinarta 2019). Understanding these dynamics is crucial to maximizing AI's potential while ensuring guest satisfaction and brand loyalty. To address these issues, this study is guided by the following research questions:

1. How does AI-driven personalisation influence customer satisfaction and loyalty in hospitality and tourism?
    - AI-driven personalisation enables tailored recommendations based on customer behaviour and preferences, significantly improving service quality and satisfaction (Vashishth et al. 2025a, Patil 2024, Li et al. 2023). Studies indicate that personalised services enhance guest engagement, leading to increased trust and repeat visits (Tussyadiah 2020). However, the effectiveness of AI personalisation depends on customer perceptions of data security and algorithmic transparency (Nwanna et al. 2025, Nama 2021, Kabadayi et al. 2019).
2. What role do ethical AI, AI transparency, and AI scepticism play in shaping trust in AI-enabled services?
    - Trust is a critical determinant of AI adoption, as customers often express concerns about privacy, data handling, and decision-making transparency (Oprea et al. 2024, Rane et al. 2024a, Buhalis and Sinarta 2019). Research suggests that businesses must implement ethical AI practices and clear communication strategies to foster consumer trust (Nadeem 2025). Additionally, AI scepticism affects the extent to which consumers rely on AI recommendations (Rajaobelina et al. 2021).
3. How does trust in AI affect AI adoption in hospitality services?
    - AI adoption is influenced by customer trust, which is shaped by factors such as perceived fairness, reliability, and transparency in

AI-driven services (Kim et al. 2024). A lack of trust can hinder adoption, while positive experiences with AI can reinforce consumer confidence in AI-driven personalisation (Alam et al. 2024, Pillai and Sivathanu 2020, Bowen and Morosan 2018).

4. How does AI adoption contribute to customer satisfaction and customer experience?
    - AI adoption leads to enhanced service efficiency, seamless transactions, and improved decision-making in hospitality services (Gursoy 2025, Collins 2024, Gretzel 2021). However, the balance between automation and human interaction remains crucial, as an over-reliance on AI may reduce emotional engagement with customers (Bhatti and Alawad 2023, Ivanov and Webster 2019).
5. How do customer satisfaction and customer experience impact customer loyalty?
    - Research suggests that positive AI-driven experiences result in higher customer satisfaction, which in turn fosters loyalty (Sardesai et al. 2024, HA 2021, Tussyadiah 2020). However, AI must be seamlessly integrated into service processes to maintain a high-quality guest experience.

This research contributes to the growing literature by providing insights into the interplay between AI-enabled personalisation, trust, and customer engagement. It emphasises the importance of ethical AI implementation, balancing automation with human touch, and fostering transparency to enhance guest experiences in the tourism and hospitality sectors.

LITERATURE REVIEW

AI-Driven Personalisation in Tourism

Artificial Intelligence (AI) is reshaping the tourism industry by offering hyper-personalised travel experiences and promoting sustainable practices (Solakis et al. 2024, Buhalis and Sinarta 2019). AI-driven tools such as chatbots, virtual assistants, and recommendation engines leverage machine learning algorithms to process vast datasets and generate tailored recommendations (Abas et al. 2024, Harun et al. 2022, Samala et al. 2022). These tools not only enhance user convenience but also facilitate sustainability efforts by guiding travellers toward eco-friendly choices, such as selecting green-certified accommodations, choosing low-carbon transportation modes, and visiting less crowded destinations (Kaakandikar et al. 2025, Chandel and Arora 2025, Majid et al. 2023, Pencarelli 2020).

However, the effectiveness of AI-driven personalisation in tourism is contingent upon its ability to align with consumer values and deliver consistently relevant outcomes (Gün 2025, Buhalis and Sinarta 2019). When AI-generated recommendations fail to align with user expectations, especially concerning sustainability, consumers may perceive them as inconvenient or even counterproductive

(Mogaji and Jain 2024). Therefore, AI systems must integrate not only technical advancements but also behavioural and ethical considerations to ensure meaningful engagement.

Factors Influencing AI Adoption

The adoption of AI in tourism is influenced by technological, psychological, and ethical factors. The Technology Acceptance Model (TAM) (Davis 1989) posits that perceived usefulness and ease of use are critical determinants of AI acceptance. Tourists evaluate AI-based recommendations based on their convenience, relevance, and security (Doğan and Niyet 2024, Song and He 2023). Furthermore, real-time data integration enhances the perceived value of AI recommendations, making them more appealing to travellers (Gretzel et al. 2015).

Recent research on AI adoption highlights the trade-offs between transparency, performance, and environmental sustainability. Ioku et al. (2024) found that consumers prioritise transparency over performance when selecting AI systems but often value performance over environmental sustainability (Hao and Demir 2024). This suggests that while users demand ethical AI solutions, practical functionality remains paramount. Additionally, environmentally conscious individuals tend to prioritise sustainability when adopting AI recommendations, aligning with broader trends in eco-conscious consumer behaviour.

Trust plays a pivotal role in AI adoption, particularly regarding transparency and fairness. Studies emphasize that scepticism about AI's reliability and fairness remains a significant barrier (Choung et al. 2023, Tussyadiah 2020). Ethical AI design, including explainable algorithms and transparent data practices, can mitigate these concerns and improve adoption rates (Wong et al. 2024, Huang and Rust 2018). Ensuring that AI systems provide unbiased, ethical recommendations can significantly enhance user trust and engagement.

Environmental Values and AI Adoption

Environmental consciousness influences AI adoption in tourism, particularly in sustainable travel solutions. Tourists with strong pro-environmental values are more likely to engage with AI-driven sustainability initiatives (Erdmann and Toro-Dupouy 2025, Miceli et al. 2021). Sharmin et al. (2020) further demonstrated that environmental sustainability awareness (ESA) plays a crucial role in encouraging environmentally sustainable behaviour (ESB). Using value-belief-norm (VBN) theory and the new ecological paradigm (NEP), their study highlighted the significance of awareness in driving pro-environmental tourism behaviours.

However, a challenge in AI-driven sustainability is the risk of "greenwashing"—superficial environmental claims that lack substantive impact. de Freitas Netto et al. (2020) warned that if AI systems promote "green" options without verifiable benefits, they may alienate eco-conscious users. Ioannou et al. (2023) suggested that businesses with strong sustainability reputations can mitigate the negative effects of perceived greenwashing by ensuring that AI-driven recommendations are credible and supported by actionable data (Santos et al. 2024).

Evolution of AI in Tourism and Hospitality

AI has undergone significant evolution over decades, with applications extending into various industries, including tourism. Initially, AI focused on logic-based problem-solving (Polo-Peña et al. 2025, Knani et al. 2022), but advancements in machine learning and deep learning have expanded its capabilities (Saydam et al. 2022). In tourism, AI applications include automated booking systems, smart tourism management, and AI-powered virtual travel assistants (Cain and Webster 2023). Recent developments in deep learning have enabled AI to analyse complex consumer preferences, enhancing its personalisation potential (Sharma 2024, Huang and Rust 2018).

Applications of AI in Hospitality and Tourism

The hospitality and tourism sectors have widely adopted AI-driven innovations, including augmented reality (AR), virtual reality (VR), blockchain, and robotics (Fatema et al. 2024, Rane et al. 2023). AI robots, equipped with motion control and data-processing capabilities, assist with hotel services such as front-desk operations, housekeeping, and food delivery (Ivanov et al. 2019, Li et al. 2023). AI-powered recommendation engines analyse past travel behaviours to curate personalised itineraries, while AR and VR provide immersive pre-travel experiences, enabling tourists to explore destinations virtually before booking trips (Tuo et al. 2025, Tuomi et al. 2021, Jeong and Shin 2020).

Role of AI Chatbots in Enhancing Customer Loyalty

AI chatbots are increasingly used in hospitality to enhance customer engagement and loyalty. These chatbots employ natural language processing to simulate human-like interactions, providing round-the-clock support and personalised recommendations (Vashishth et al. 2024). Research indicates that AI chatbots improve customer satisfaction through rapid response times and efficient problem-solving (Vashishth et al. 2025b, Rani et al. 2024). However, their effectiveness is contingent on perceived intelligence, warmth, and social presence.

Despite their advantages, AI chatbots present challenges, including concerns about emotional intelligence and privacy. Customers expect chatbots to understand their needs while maintaining data security (Waladi et al. 2024). Additionally, overly human-like chatbots can create a sense of "uncanny valley," reducing trust (Rajaobelina et al. 2021). Studies suggest that a hybrid model integrating AI with human service can maximize customer loyalty while minimizing potential drawbacks (Sağbaş and Aydogan 2025, Inavolu 2024).

AI Chatbots and Hotel Customer Loyalty

Hotels are leveraging AI chatbots to enhance guest experiences and foster loyalty. Research shows that high-quality chatbot interactions significantly influence hotel customer satisfaction (Gupta and Pareek 2024). Xia and Shannon (2025) found that chatbot warmth and competence play a crucial role in customer expectations of service quality. However, negative perceptions, such as "creepiness" or privacy concerns, can hinder chatbot adoption (Malkomsen 2024).

Moreover, the effectiveness of AI chatbots varies based on hotel segments and customer demographics. Hanji et al. (2024) reported that luxury hotel guests are more likely to embrace AI chatbots compared to mid-range travellers. Similarly, Li et al. (2023) highlighted that customer innovativeness and involvement influence chatbot-driven loyalty, suggesting that hotels must tailor AI strategies to different consumer groups.

RESEARCH OBJECTIVES

This study aims to examine the role of AI-driven personalisation in hospitality and tourism, focusing on customer satisfaction, trust, AI adoption, and customer loyalty. Specifically, the objectives are:

1. To analyse the impact of AI-driven personalisation on customer satisfaction and loyalty.
2. To investigate the role of ethical AI, AI transparency, and AI scepticism in shaping trust in AI-powered services.
3. To evaluate the influence of trust in AI on AI adoption in the hospitality sector.
4. To examine how AI adoption contributes to customer satisfaction and overall customer experience.
5. To assess the relationship between customer satisfaction, customer experience, and customer loyalty.

HYPOTHESES DEVELOPMENT

AI-Driven Personalisation and Customer Satisfaction

AI-driven personalisation enhances customer satisfaction by delivering tailored recommendations based on consumer behaviour and preferences (Li et al. 2023). Studies indicate that personalised services increase engagement, trust, and brand loyalty, particularly when AI successfully predicts customer needs (Tussyadiah 2020). However, the effectiveness of AI personalisation depends on customer perceptions of data security and algorithmic transparency (Kabadayi et al. 2019).

H_1: *AI-driven personalisation positively influences customer satisfaction in hospitality and tourism.*

H_2: *Customer satisfaction with AI-driven personalisation positively impacts customer loyalty.*

Trust and AI-Enabled Service Adoption

Trust plays a critical role in shaping consumer attitudes toward AI-driven services. Guests may hesitate to adopt AI solutions due to concerns about data privacy, algorithmic bias, and decision-making transparency (Buhalis and Sinarta 2019). Research highlights that trust can be fostered through ethical AI practices, clear communication, and seamless integration of AI with human service (Miceli et al. 2021). Additionally, AI should not entirely replace human interaction, as emotional engagement remains essential for customer experience (Bowen and Morosan 2018).

***H_3*:** *Ethical AI practices positively influence consumer trust in AI-driven personalisation.*

***H_4*:** *AI transparency positively influences consumer trust in AI-driven personalisation.*

***H_5*:** *AI scepticism negatively influences consumer trust in AI-driven personalisation.*

***H_6*:** *Trust in AI-enabled services positively influences AI adoption in hospitality and tourism.*

Trust in AI and Customer Satisfaction

Trust in AI is a crucial factor in shaping customer satisfaction. When AI-powered services are perceived as reliable, fair, and transparent, customers are more likely to develop positive experiences, leading to higher satisfaction (Coelho and Imamović 2025, Rane et al. 2024b).

***H_7*:** *Trust in AI positively influences customer satisfaction in hospitality services.*

Customer Loyalty and Experience

Customer experience and satisfaction are fundamental in driving customer loyalty. AI-driven services should not only meet customer expectations but also provide value beyond automation (Rane et al. 2024c, Venkateswaran et al. 2024). Ensuring high-quality service and engagement through AI can lead to long-term customer retention.

***H_8*:** *Customer loyalty positively impacts positive customer experience.*

RESEARCH METHODOLOGY

This study employs a quantitative research design, utilizing survey-based data collection and statistical analysis to investigate the relationships between AI-driven personalisation, customer satisfaction, trust, AI scepticism, transparency, and adoption in the hospitality and tourism industry. A cross-sectional approach is adopted to gather responses from a diverse set of hospitality customers, providing insights into their experiences with AI-powered services (Zhou et al. 2024, Manzoor et al. 2024).

Research Framework and Description

Framework presented in Figure 6.1 highlights importance of trust, transparency, and a balanced AI-human interaction, while also accounting for the moderating effect of AI scepticism. The framework serves as foundation for hypothesis testing and empirical validation in AI-driven guest experience engineering. The above framework is basically structured into four key components:

- Antecedents (Independent Variables)—AI driven personalisation, Trust in AI-Enabled Services, Ethical AI Practices and Transparency, AI-Human Interaction
- Moderator—AI Scepticism.
- Mediator—Customer Satisfaction with AI Services.
- Dependent Variable—Customer Loyalty.

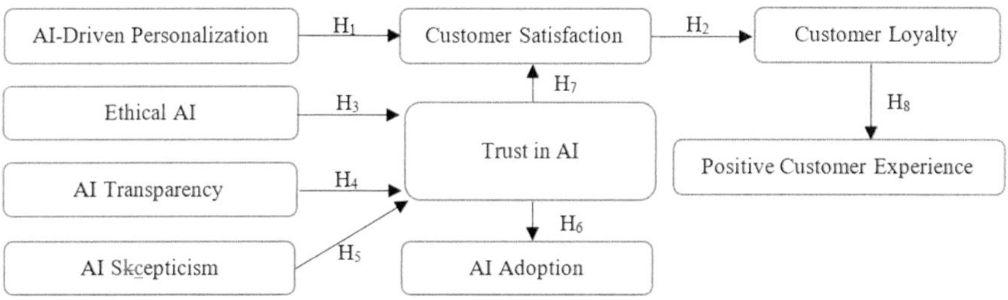

Figure 6.1 Research framework (*Source:* Author's Compilation).

POPULATION AND SAMPLE

- ***Population:*** The target population consists of hospitality and tourism customers who have interacted with AI-driven services, such as personalised hotel recommendations, AI-powered chatbots, virtual concierges, and automated booking systems (Vashishth et al. 2024, Buhalis and Sinarta 2019). The study focuses on customers in hotels, airlines, and online travel agencies, as these sectors actively implement AI-based personalisation (Tussyadiah 2020).
- ***Sampling Technique*:** A purposive sampling technique is used to ensure that respondents have prior exposure to AI-driven personalisation services in tourism (Ionescu and Sârbu 2024, Saunders and Darabi 2024). A quota sampling approach ensures diverse representation across customer demographics, including age, travel frequency, and technology affinity.
- ***Sample Size Determination*:** The sample size is determined using Cohen's (1998) power analysis to achieve adequate statistical power for hypothesis testing. Based on prior studies (e.g. Kabadayi et al. 2019), a minimum of 600–700 respondents are required to ensure robust analysis through Structural Equation Modelling (SEM).

Research Instrument: A structured self-administered questionnaire is developed, including validated scales from prior studies. The questionnaire consists of five sections:

- Demographic information (age, gender, education, travel frequency, AI familiarity).
- AI-driven Personalisation (IV)—Adapted from Li et al. (2023).
- Customer Satisfaction and Loyalty (DV)—Based on Tussyadiah (2020).
- Trust and AI Transparency (MV)—Using measures from Buhalis and Sinarta (2019).
- AI Scepticism (Moderator)—Based on Rajaobelina et al. (2021).

All responses use a 5-point Likert scale (1 = Strongly Disagree, 5 = Strongly Agree) to ensure measurement precision (Edmondson 2005).

Data Collection: The data is collected through online surveys distributed via email and social media platforms, targeting hospitality customers worldwide. To ensure high-quality responses, screening questions filter out respondents who have never used AI-driven personalisation services.

Data Analysis

Table 6.1 presents the preliminary data screening process conducted before performing advanced statistical analyses, which helps to understand handling of missing data effectively, to ensure the reliability and validity of subsequent analyses.

Table 6.1 Data analyses

Preliminary Analysis	
Missing Data Handling	Listwise deletion for missing values (Hair et al. 2019)
Hypothesis Testing: The study employs Structural Equation Modelling (SEM), a powerful tool for path analysis and hypothesis testing (Lowry and Gaskin 2014).	
Exploratory Factor Analysis (EFA)	Identifies latent constructs and ensures validity.
Confirmatory Factor Analysis (CFA)	Tests model fit indices (CFI, TLI, and RMSEA).
Path Analysis	Assesses direct and indirect relationships among variables.
Moderation Analysis	Examines the impact of AI scepticism on AI trust and adoption.
Reliability and Validity Tests	
Cronbach's Alpha	Ensures internal consistency (Nunnally 1978)
Composite Reliability (CR)	Evaluates construct reliability
Convergent and Discriminant Validity	Measured via Average Variance Extracted (AVE) and Fornell-Larcker criterion (Fornell and Larcker 1981)

Table 6.2 Respondents' profile (N = 680)

Demographic Variables (Personal and Socio-Economic)			
Variable	**Categories**	**Frequency (N)**	**Percent (%)**
Gender	Male	295	43.40
	Female	320	47.10
	Others	45	6.60
	Prefer not to say	20	2.90
Age Group	18–25 years	180	26.50
	26–35 years	240	35.30
	36–45 years	130	19.10
	46–55 years	80	11.80
	56 years and above	50	7.30
Marital Status	Single	350	51.50
	Married	180	26.50
	Married with children	80	11.80
	Divorced	30	4.40
	Widowed	20	2.90
	Separated due to work	20	2.90
Educational Status	High School or below	80	11.80
	Bachelor's degree	380	55.90
	Master's degree	140	20.60
	Doctorate	40	5.80
	Others (Vocational/Diploma)	40	5.90
Occupational Status	Student	100	14.70
	Government employee	70	10.30
	Private sector employee	260	38.20
	Business owner/Entrepreneur	90	13.20
	Freelancer	70	10.30
	Tourism and hospitality professional	40	5.90
	Tech/AI industry professional	30	4.40
	Retired	20	2.90
Monthly Income (INR)	Less than ₹20,000	80	11.80
	₹20,000–₹50,000	150	22.10
	₹50,001–₹1,00,000	220	32.40
	₹1,00,001–₹2,00,000	120	17.60
	Above ₹2,00,000	110	16.20
Residential Location	Urban	460	67.60
	Suburban	150	22.10
	Rural	70	10.30
Nationality	Domestic Tourist	450	66.20
	International Tourist	230	33.80

Source: Frequency distribution analysis output

Results and Discussion

This section outlines statistical outputs and the key findings derived from the structural equation modelling analysis, highlighting the interaction between AI-related tourism dimensions, customer satisfaction, loyalty, and positive customer experiences in tourism. The section also presents the validation of constructs, hypothesis testing, and mediation effects, along with model fit and predictive relevance. The discussion focuses on supported and unsupported relationships, offering insights into AI-driven personalisation, ethical AI, AI scepticism, AI adoption, trust in AI, etc.

Table 6.2 presents the demographic distribution of the 680 study participants, capturing both personal and socio-economic characteristics. In terms of gender, 295 participants (43.4%) identified as male, 320 (47.1%) as female, 45 (6.6%) as others, and 20 (2.9%) chose not to disclose. The age group with the highest representation was 26–35 years with 240 respondents (35.3%), followed by 180 individuals (26.5%) aged 18–25 years. Other age groups included 130 (19.1%) aged 36–45, 80 (11.8%) aged 46–55, and 50 (7.3%) aged 56 and above.

Marital status data revealed that a majority, 350 (51.5%), were single, while 180 (26.5%) were married and 80 (11.8%) were married with children. Smaller groups included 30 divorced (4.4%), 20 widowed (2.9%), and 20 separated due to work (2.9%). Educational qualifications were relatively high: 380 participants (55.9%) held a bachelor's degree, 140 (20.6%) a master's, 40 (5.8%) a doctorate, 80 (11.8%) had high school education or below, and 40 (5.9%) had vocational or diploma-level education.

Occupational status showed a diverse mix, with the private sector employing the largest share at 260 individuals (38.2%). Other categories included 100 students (14.7%), 90 business owners/entrepreneurs (13.2%), 70 government employees (10.3%), 70 freelancers (10.3%), 40 tourism and hospitality professionals (5.9%), 30 tech/AI professionals (4.4%), and 20 retirees (2.9%). Regarding monthly income, 220 (32.4%) earned between ₹50,001–₹1,00,000, 150 (22.1%) earned ₹20,000–₹50,000, 120 (17.6%) earned ₹1,00,001–₹2,00,000, 110 (16.2%) earned above ₹2,00,000, and 80 (11.8%) earned below ₹20,000.

A significant portion, 460 respondents (67.6%), lived in urban areas, with 150 (22.1%) residing in suburban and 70 (10.3%) in rural areas. Lastly, the sample comprised 450 domestic tourists (66.2%) and 230 international tourists (33.8%), capturing both local and global perspectives for the study. This comprehensive demographic representation adds depth and diversity to the research insights on AI-driven personalisation in tourism.

The evaluation of the measurement model presented in Table 6.3 revealed a high degree of reliability and validity across all the constructs, affirming the robustness of the instrument. Convergent validity is strongly established, with all factor loadings (FL) exceeding the minimum recommended threshold of 0.70, ranging from 0.811 to 0.907, underlining substantial correlation between observed variables and their underlying constructs. The Average Variance Extracted (AVE) values for all constructs surpass the 0.50 benchmark, ranging from 0.704 to 0.791, indicating that each construct captures more than half of the variance in its indicators, further confirming convergent validity.

Table 6.3 Construct reliability and validity assessment

Construct	Code	Items	VIF	FL	α	CR	AVE
AI-Driven Personalisation (AIP)	AIP1	AI-based recommendations align with my preferences.	1.932	0.847	0.871	0.879	0.710
	AIP2	AI-driven personalisation advances my experience.	2.054	0.862			
	AIP3	AI personalisation helps me make better choices.	1.844	0.878			
Customer Satisfaction (SAT)	SAT1	I am satisfied with AI-driven service recommendations.	2.111	0.875	0.868	0.869	0.791
	SAT2	AI-based services meet my expectations.	2.288	0.893			
	SAT3	I have a positive attitude toward AI-assisted services.	2.500	0.900			
Customer Loyalty (LOY)	LOY1	I will recommend AI-based services to others.	2.572	0.907	0.858	0.859	0.779
	LOY2	I will continue using AI-driven services.	2.212	0.883			
	LOY3	AI-based services are my preferred choice.	1.949	0.857			
Ethical AI (ETH)	ETH1	AI systems provide fair recommendations.	2.004	0.841	0.832	0.846	0.737
	ETH2	AI respects user privacy and data security.	1.893	0.876			
AI Transparency (AIT)	AIT1	AI explains why it makes certain recommendations.	2.013	0.851	0.829	0.841	0.721
	AIT2	AI-based services provide clear and understandable information.	2.007	0.846			
AI Scepticism (AIS)	AIS1	I am sceptical about AI-driven recommendations.	1.768	0.811	0.817	0.826	0.704
	AIS2	I trust human recommendations more than AI-based ones.	1.872	0.868			
Trust in AI (TIA)	TIA1	AI-driven services are reliable.	2.206	0.889	0.866	0.871	0.768
	TIA2	I trust AI-based personalisation.	2.149	0.864			
AI Adoption (AIA)	AIA1	I am likely to adopt AI-driven services.	1.912	0.836	0.841	0.852	0.724
	AIA2	AI-based recommendations influence my decisions.	1.988	0.869			
Positive Customer Experience (PCE)	PCE1	AI-driven services enhance my overall experience.	2.051	0.881	0.861	0.867	0.757
	PCE2	AI personalisation makes my interaction seamless.	2.077	0.891			

Source: **Test output (Note:** VIF = Variance Inflation Factor; FL = Factor Loadings; α = Cronbach's Alpha; CR = Composite Reliability; AVE = Average Variance Extracted. Sources will be updated accordingly.)

In terms of internal consistency, Cronbach's Alpha (α) values are consistently above 0.80, specifically ranging between 0.817 (AI Scepticism) and 0.871 (AI-Driven Personalisation), exceeding the conventional threshold of 0.70 and suggesting strong reliability. Similarly, Composite Reliability (CR) values fall within the range of 0.826 to 0.879, indicating excellent internal consistency across all measurement scales. Furthermore, all Variance Inflation Factor (VIF) values were below the critical value of 3.3, ranging from 1.768 to 2.572, suggesting the absence of multicollinearity among the indicators and supporting the model's statistical integrity. Notably, constructs such as Customer Satisfaction (SAT) and Customer Loyalty (LOY) demonstrated particularly high factor loadings up to 0.900 and 0.907 respectively, signalling the strength of AI's influence on user attitudes and behavioural intentions in the tourism sector.

Overall, the results substantiate that the model is both psychometrically sound and theoretically grounded, and each construct exhibits high reliability, internal consistency, and adequate convergent validity, thereby validating the appropriateness of the proposed measurement framework in the context of AI-enabled personalisation in tourism experiences.

Table 6.4 Discriminant validity for the Fornell–Larcker criterion

Construct	AIP	SAT	LOY	ETH	AIT	AIS	TIA	AIA	PCE
AI-Driven Personalisation (AIP)	**0.842**								
Customer Satisfaction (SAT)	0.711	**0.889**							
Customer Loyalty (LOY)	0.695	0.742	**0.883**						
Ethical AI (ETH)	0.684	0.728	0.718	**0.858**					
AI Transparency (AIT)	0.672	0.715	0.682	0.766	**0.849**				
AI Scepticism (AIS)	0.593	0.609	0.577	0.638	0.643	**0.839**			
Trust in AI (TIA)	0.707	0.774	0.733	0.722	0.704	0.597	**0.876**		
AI Adoption (AIA)	0.749	0.733	0.745	0.701	0.719	0.588	0.768	**0.851**	
Positive Customer Experience (PCE)	0.723	0.768	0.776	0.738	0.726	0.603	0.754	0.781	**0.870**

Source: Test output

Discriminant validity statistics presented in Table 6.4, calculated using the Fornell–Larcker criterion, confirm that all constructs of the model are distinct from one another, supporting the model's structural robustness. In this criterion, discriminant validity is established when the square root of the Average Variance Extracted (AVE) for each construct is found to be greater than its correlations with other constructs. In this model, all diagonal values, ranging from 0.839 (AI Scepticism) to 0.889 (Customer Satisfaction), were found higher than the corresponding inter-construct correlations, which satisfies the necessary condition for discriminant validity.

For instance, the square root of AVE for AI-Driven Personalisation (AIP) is 0.842, which is greater than its correlations with SAT (0.711), LOY (0.695), and other constructs, indicating that AIP shares more variance with its own indicators than with those of any other construct. A similar pattern is observed for other variables, such as Trust in AI (TIA) with a diagonal value of 0.876, higher than its highest inter-construct correlation (0.774 with SAT), and Positive Customer Experience (PCE), which shows a strong AVE square root of 0.870, exceeding all its correlations with constructs like LOY (0.776) and AIP (0.723).

The empirical results affirm that each construct captures a unique aspect of the AI-personalisation experience in tourism, free from excessive overlap with other constructs. This strongly supports the model's discriminant validity and ensures that the constructs reflect conceptually distinct dimensions such as trust, ethics, satisfaction, and scepticism in relation to AI technologies in customer experience design.

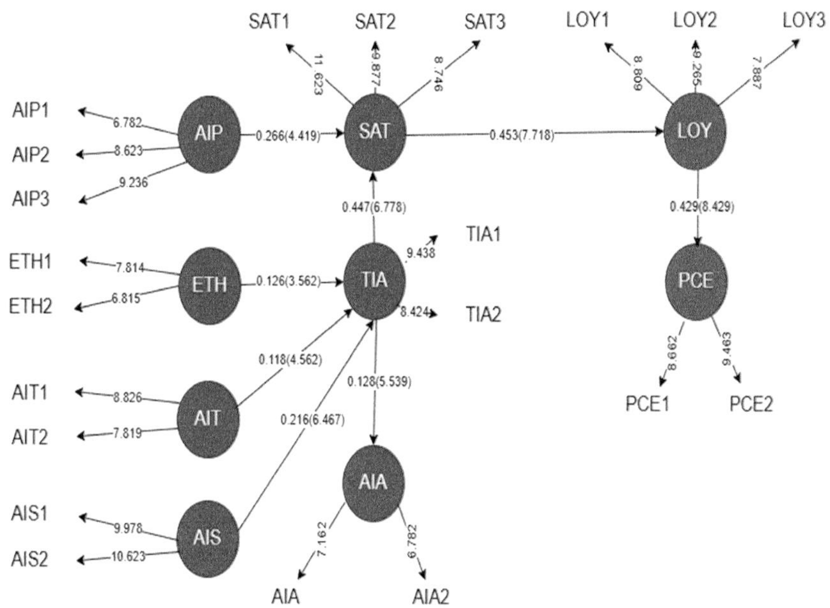

Figure 6.2 SEM: Standardized path coefficients and measurement loadings (*Source:* Statistical Results of the Research Framework).

Standardized path coefficients and measurement loadings presented in Figure 6.2 and hypothesis testing results presented in Table 6.5 offer compelling evidence in support of the conceptual model proposed for AI-driven personalisation in the hospitality and tourism industry. All eight hypotheses are statistically significant with p-values of 0.000, indicating strong empirical support across all hypothesized relationships. H_1 is supported with a standardized beta value of 0.266 and a t-statistic of 4.419, confirming that AI-driven personalisation (AIP) significantly and positively impacts customer satisfaction (SAT) in hospitality and tourism. This finding underscores the role of tailored AI experiences in enhancing guest satisfaction. H_2 and H_3 revealed that both ethical AI practices (ETH → TIA, $\beta = 0.126$, $t = 3.562$) and AI transparency (AIT → TIA, $\beta = 0.118$, $t = 3.562$) positively influence trust in AI (TIA). These relationships reinforce the importance of fairness, clarity, and openness in building consumer trust in AI-driven systems.

Table 6.5 Results of the hypothesis testing.

Hypothesis	Relationships	Original sample (β)	T-statistics	P-values	Decision
H_1	AIP → SAT	0.266	4.419	0.000	Support
H_2	ETH → TIA	0.126	3.562	0.000	Support
H_3	AIT → TIA	0.118	3.562	0.000	Support
H_4	AIS → TIA	0.216	6.467	0.000	Support
H_5	TIA → SAT	0.447	6.778	0.000	Support
H_6	TIA → AIA	0.128	5.539	0.000	Support
H_7	SAT → LOY	0.453	7.718	0.000	Support
H_8	LOY → PCE	0.429	8.429	0.000	Support

Source: Figure 6.2 and test output, hypotheses interpretation

In H_4, AI scepticism (AIS) showed a significant positive relationship with Trust in AI (TIA) ($\beta = 0.216$, $t = 6.467$). Although scepticism might typically suggest reluctance, this result could indicate that a critical evaluation of AI systems may lead to more informed and confident adoption decisions when trust is built. H_5 and H_6 confirmed that trust in AI (TIA) plays a central mediating role in the model. TIA significantly influences both customer satisfaction ($\beta = 0.447$, $t = 6.778$) and AI adoption ($\beta = 0.128$, $t = 5.539$), emphasizing that fostering user trust is crucial for both satisfaction and continued engagement with AI technologies. H_7 affirms that customer satisfaction has a strong and significant impact on loyalty (SAT → LOY, $\beta = 0.453$, $t = 7.718$), aligning with well-established loyalty models in service marketing. Lastly, H_8 demonstrated that customer loyalty (LOY) positively influences positive customer experience (LOY → PCE, $\beta = 0.429$, $t = 8.429$), reinforcing that loyal customers are more likely to perceive and report enhanced service experiences.

Overall, results validated the interconnected roles of AI personalisation, ethical and transparent practices, and trust in shaping satisfaction, loyalty, and experience in the hospitality and tourism domain.

Mediation analysis values presented in Table 6.6 reveal significant multi-path relationships that enhance the understanding of AI-driven personalisation's influence on customer experiences in tourism and hospitality. The pathway AIP → SAT → LOY → PCE ($\beta = 0.052$, $p = 0.002$) confirms that customer satisfaction and loyalty play a pivotal mediating role in translating AI personalisation into enriched customer experiences. Similarly, the indirect effects of ethical AI, transparency, and scepticism through trust in AI, satisfaction, and loyalty were all statistically significant. Specifically, ETH → TIA → SAT → LOY → PCE ($\beta = 0.011$, $p = 0.033$), AIT → TIA → SAT → LOY → PCE ($\beta = 0.010$, $p = 0.039$), and AIS → TIA → SAT → LOY → PCE ($\beta = 0.019$, $p = 0.003$) illustrate how cognitive trust mechanisms amplify the customer journey from AI ethics and clarity towards loyalty-driven experiences.

Table 6.6 Mediation effects

Mediation Path	Original Sample (β)	T-statistics	P-values	Decision
AIP → SAT → LOY → PCE	0.052	3.102	0.002	Yes
ETH → TIA → SAT → LOY → PCE	0.011	2.141	0.033	Yes
AIT → TIA → SAT → LOY → PCE	0.010	2.063	0.039	Yes
AIS → TIA → SAT → LOY → PCE	0.019	2.998	0.003	Yes
ETH → TIA → AIA	0.016	2.421	0.016	Yes
AIT → TIA → AIA	0.015	2.352	0.019	Yes
AIS → TIA → AIA	0.028	3.887	0.000	Yes

Source: Figure 6.2 and test output

Moreover, mediation through AI adoption was also supported. The paths ETH → TIA → AIA ($\beta = 0.016$, $p = 0.016$), AIT → TIA → AIA ($\beta = 0.015$, $p = 0.019$), and AIS → TIA → AIA ($\beta = 0.028$, $p = 0.000$) reinforce that trust in AI is a key driver for the adoption of AI, significantly influenced by users' perceptions of AI fairness, transparency, and scepticism. These findings affirm the critical mediating role of trust and satisfaction in AI engagement, shaping both emotional and behavioural responses in the tourism and hospitality context.

Table 6.7 Quality criteria

Construct	R-square	R-square adjusted
TIA	0.587	0.582
LOY	0.694	0.692
SAT	0.765	0.758
PCE	0.694	0.691

Source: Test output

The explanatory capability of the structural model was assessed using the coefficient of determination (R^2), as recommended by Bagozzi and Yi (2012). From Quality criteria measurements for different constructs presented in Table 6.7, it was observed that all R^2 values exceeded the minimum benchmark of 0.10, thereby confirming the model's predictive adequacy. Notably, the constructs Customer Satisfaction (SAT) and Customer Loyalty (LOY) demonstrated high explanatory power with R^2 values of 0.765 and 0.694, respectively. According to the classification proposed by Henseler et al. (2009), these values reflect substantial (≥ 0.75) and moderate-to-substantial (≥ 0.50) predictive accuracy.

Additionally, the construct Trust in AI (TIA) exhibited a moderate R^2 value of 0.587, while Positive Customer Experience (PCE) also achieved a moderately high R^2 of 0.694. These results collectively indicate that the model is capable of explaining a significant proportion of variance across key dependent constructs. The adjusted R^2 values for all constructs closely align with the R^2 values, further validating the model's internal consistency and robustness.

Table 6.8 Model fit

Criteria	Saturated Model	Estimated Model
SRMR	0.041	0.056

Source: Test output

The overall model fit was assessed using Standardized Root Mean Square Residual (SRMR), a widely accepted indicator for evaluating structural model fitness. SRMR recommended threshold of 0.08 is generally considered indicative of a good model fit. According to model fit values of Table 6.8, SRMR value is 0.041 for the saturated model and 0.056 for the estimated model, both comfortably within acceptable limits. These results suggest that the discrepancy between the observed and predicted correlations is minimal, indicating a strong model-data fit. Therefore, the structural model can be considered well-specified and appropriate.

Table 6.9 Construct prediction summary

Construct	Q^2_predict	RMSE	MAE
TIA	0.684	0.531	0.379
LOY	0.706	0.549	0.409
SAT	0.738	0.520	0.360
PCE	0.667	0.507	0.351

Source: Test output

The predictive relevance of the model was evaluated using Q^2_predict, Root Mean Square Error (RMSE), and Mean Absolute Error (MAE). In accordance with Construct Prediction Summary presented in Table 6.9, all constructs, namely TIA (0.684), LOY (0.706), SAT (0.738), and PCE (0.667), have exhibited positive Q^2_predict values, which confirms model's strong predictive relevance, as per the guidelines by Hair et al. (2019). The RMSE values for the constructs range from 0.507 (PCE) to 0.549 (LOY), while MAE values range between 0.351 (PCE) and 0.409 (LOY). The relatively low RMSE and MAE values across all constructs indicate high predictive accuracy and minimal error in the model's estimations. Hence, it can be concluded that the model demonstrates excellent predictive performance across the key outcome variables.

DISCUSSION AND CONCLUSIONS

Leveraging Structural Equation Modelling (SEM), the results highlight the strong theoretical foundation and statistical robustness of the proposed model, validated through high construct reliability, convergent and discriminant validity, as well as significant mediation effects and hypothesis testing outcomes. The demographic profile of the 680 participants revealed a balanced representation across gender, age groups, marital status, and income groups, offering a robust and inclusive sample base. The high representation of individuals aged 26–35 (35.3%) and urban residents (67.6%) reflects a population segment more likely to interact with digital and AI-driven tourism platforms. The prevalence of bachelor's degree holders (55.9%) and private sector employees (38.2%) suggested an audience with both digital literacy and purchasing power, which are critical for adopting AI solutions in tourism. Further, discriminant validity, assessed using the Fornell–Larcker criterion, affirmed that each construct is statistically distinct from the others. The square roots of AVE values for all constructs were greater than their respective inter-construct correlations, confirming that the constructs uniquely capture specific elements of AI-related experiences in tourism. For example, Trust in AI (TIA) had a square root AVE of 0.876, higher than any of its correlations with other constructs, thereby underlining its conceptual uniqueness.

The hypothesis testing results offered strong support for the proposed structural relationships. All eight hypotheses were validated with high t-statistics and statistically significant p-values ($p < 0.001$), reinforcing the theoretical assumptions.

- Hypothesis H_1 confirmed that AI-driven personalisation (AIP) positively impacts customer satisfaction (SAT), highlighting the importance of tailored experiences in improving user engagement (Wijethilak et al. 2025, Balasubramanian 2024).
- Hypotheses H_2 and H_3 demonstrated that both ethical considerations and transparency in AI significantly enhance trust in AI (TIA), underscoring the need for fairness, accountability, and openness in AI systems (Farinu 2025, Liebherr et al. 2025, Emma 2024).
- Hypothesis H_4, while initially counterintuitive, showed that AI scepticism (AIS) positively influences trust (Daly et al. 2025, Oprea et al. 2024, Zerilli et al. 2022). This suggests that informed scepticism may drive users to critically assess AI systems and develop trust based on reasoned acceptance rather than blind faith.
- Hypotheses H_5 and H_6 underscored the mediating role of trust in AI. Trust strongly impacts customer satisfaction and the likelihood of AI adoption (Chotisarn and Phuthong 2025, Nguyen et al. 2023, Shi et al. 2021), highlighting that without trust, the utility of AI personalisation remains limited.
- Lastly, hypotheses H_7 and H_8 confirmed the established service marketing theories (Mascarenhas et al. 2006, Cetin 2020) by showing that satisfaction leads to loyalty, and loyalty significantly enhances the customer experience (Huddin et al. 2024, Ifekanandu et al. 2023).

The mediation analysis further enriched the findings by demonstrating the significant indirect effects across multiple pathways. The mediated relationship from AIP → SAT → LOY → PCE confirmed that both satisfaction and loyalty are essential intermediaries between AI personalisation and a positive customer experience. This highlighted the importance of not just personalising content, but also ensuring that such personalisation fosters satisfaction and repeat engagement. Additionally, the paths from ethical AI, transparency, and scepticism to customer experience, mediated through trust, satisfaction, and loyalty, illustrated the multi-dimensional nature of trust formation and its ripple effects across the customer journey. For instance, ETH → TIA → SAT → LOY → PCE and AIT → TIA → SAT → LOY → PCE demonstrated how transparency and ethical practices indirectly enhance customer experiences through a chain of trust and emotional connection. Similarly, the pathway AIS → TIA → AIA suggested that critical users, once convinced of an AI system's reliability, may actually become strong adopters.

From a managerial perspective, the results suggest that organisations in the tourism sector must adopt a holistic AI strategy, which goes beyond technological functionality to encompass ethical standards, transparency, and trust-building mechanisms. Simply integrating AI for personalisation is insufficient; therefore, organisations must ensure that AI operates in ways that are fair, transparent, and respectful of user concerns. For researchers, the findings offer new insights into the paradoxical yet positive role of scepticism in building trust and promoting adoption. This introduces an interesting dynamic in AI acceptance models and encourages a re-examination of user attitudes in AI research.

In conclusion, this study contributes valuable empirical evidence to the discourse on AI in tourism by presenting a robust framework that links AI-driven personalisation with customer satisfaction, trust, adoption, loyalty, and ultimately, enhanced customer experiences. The validated model confirms the importance of ethical and transparent AI practices, as well as the nuanced role of scepticism, in shaping user trust and engagement. These findings not only inform AI implementation strategies in the tourism and hospitality industry but also pave the way for more human-centred AI design approaches that prioritise experience, fairness, and trust.

REFERENCES

Abas, R., Radzi, S.M. and Zahari, M.S.M. 2024. AI Adoption in Islamic tourism stakeholders' business orientation. International Journal of Business and Technology Management. 6(1): 1–14. Available at: https://myjms.mohe.gov.my/index.php/ijbtm/article/view/25862

Alam, S.S., Masukujjaman, M., Mohamed Makhbul, Z.K., Helmi Ali, M., Ahmad, I. and Al Mamun, A. 2024. Experience, trust, eWOM engagement and usage intention of AI enabled services in hospitality and tourism industry: Moderating mediating analysis. Journal of Quality Assurance in Hospitality and Tourism. 25(6): 1635–1663. https://doi.org/10.1080/1528008X.2023.2167762

Amtul, Q., Saleem, S., Mushtaq, A., Sajjad, S. and Sattar, S. 2025. Unlocking customer loyalty through the interplay of AI service quality, perceived value, and trust in hospitality industry: Mediation role of customer satisfaction. Journal of Business and Management Research. 4(1): 861–889. https://doi.org/10.62019/260pp243

Avula, M. and Sithole, T. 2024. Artificial intelligence's potential to improve operational efficiency and customer experience in the hospitality and tourism sectors. *In*: Impact of AI and Tech-Driven Solutions in Hospitality and Tourism. 351–382. IGI Global.

Balasubramanian, G. 2024. The role of AI in enhancing personalisation in ecommerce: A study on customer engagement and satisfaction. Asian and Pacific Economic Review. 17(2): 160–177.

Bagozzi, R.P. and Yi, Y. 2012. Specification, evaluation, and interpretation of structural equation models. Journal of the Academy of Marketing Science. 40(1): 8–34. https://doi.org/10.1007/s11747-011-0278-x

Bhatti, M.A. and Alawad, W.A. 2023. Technology adoption and customer loyalty in hospitality sector: Mediating role of customer satisfaction and engagement. Przestrzeń Społeczna (Social Space). 23(3): 431–459. Retrieved from: https://socialspacejournal.eu/menu-script/index.php/ssj/article/view/276

Bowen, J.T. and Morosan, C. 2018. Beware hospitality industry: The robots are coming. Worldwide Hospitality and Tourism Themes. 10(6): 726–733.

Buhalis, D. and Sinarta, Y. 2019. Real-time co-creation and nowness service: Lessons from tourism and hospitality. Journal of Travel and Tourism Marketing. 36(5): 563–582. https://doi.org/10.1080/10548408.2019.1592059

Cain, L. and Webster, C. 2023. The origin, rationale, and impacts of regulations on robots and AI in tourism and hospitality. *In*: Proceedings INNODOCT/22. International Conference on Innovation, Documentation and Education. 111–118. Editorial Universitat Politècnica de València.

Cetin, G. 2020. Experience vs quality: predicting satisfaction and loyalty in services. The Service Industries Journal. 40(15–16): 1167–1182. https://doi.org/10.1080/02642069.2020.1807005

Chandel, M. and Arora, M. 2025. Achieving the sustainable development goals through financial inclusion: A bibliometric and content analysis. Tourism and Management Studies. 21(1): 21–37. https://doi.org/10.18089/tms.20250102

Chotisarn, N. and Phuthong, T. 2025. Impact of artificial intelligence-enabled service attributes on customer satisfaction and loyalty in chain hotels: Evidence from coastal tourism destinations in western Thailand. Social Sciences and Humanities Open. 11: 101306. https://doi.org/10.1016/j.ssaho.2025.101306

Choung, H., David, P. and Ross, A. 2023. Trust in AI and its role in the acceptance of AI technologies. International Journal of Human–Computer Interaction. 39(9): 1727–1739. https://doi.org/10.1080/10447318.2022.2050543

Coelho, M.C.C.D. and Imamović, I. 2025. AI-driven personalisation in beauty retail: exploring how AI-based applications influence customer satisfaction and brand loyalty. *In*: Leveraging AI for Effective Digital Relationship Marketing. 131–162. IGI Global.

Cohen, M.P. 1998. Determining sample sizes for surveys with data analyzed by hierarchical linear models. Journal of Official Statistics. 14(3): 267.

Collins, G. 2024. Impact of digitalization on customer experience management in the hospitality industry. Hospitality and Tourism Journal. 1(1): 12. Retrieved from https://forthworthjournals.org/journals/index.php/HTJ/article/view/67

Daly, S.J., Wiewiora, A. and Hearn, G. 2025. Shifting attitudes and trust in AI: Influences on organizational AI adoption. Technological Forecasting and Social Change. 215: 124108. https://doi.org/10.1016/j.techfore.2025.124108

Davis, F.D. 1989. Technology acceptance model: TAM. Al-Suqri, MN, Al-Aufi, AS: Information Seeking Behavior and Technology Adoption. 205(219): 5.

de Freitas Netto, S.V., Sobral, M.F.F., Ribeiro, A.R.B. and Soares, G.R.D.L. 2020. Concepts and forms of greenwashing: A systematic review. Environmental Sciences Europe. 32: 1–12. https://doi.org/10.1186/s12302-020-0300-3

Doğan, S. and Niyet, İ.Z. 2024. Artificial intelligence (AI) in tourism. *In*: Future Tourism Trends Volume 2: Technology Advancement, Trends and Innovations for the Future in Tourism. 3–21. Emerald Publishing Limited. https://doi.org/10.1108/978-1-83753-970-320241001

Edmondson, D. 2005. Likert scales: A history. *In*: Proceedings of the Conference on Historical Analysis and Research in Marketing. 12: 127–133.

Emma, L. 2024. The Ethical Implications of Artificial Intelligence: A Deep Dive into Bias, Fairness, and Transparency.

Erdmann, A. and Toro-Dupouy, L. 2025. The influence of the institutional environment on AI adoption in universities: Identifying value drivers and necessary conditions. European Journal of Innovation Management. https://doi.org/10.1108/EJIM-04-2024-0407

Ernst, E., Merola, R. and Samaan, D. 2019. Economics of artificial intelligence: Implications for the future of work. IZA Journal of Labor Policy. 9(1): 1–35.

Farinu, U. 2025. Fairness, Accountability, and Transparency in AI: Ethical Challenges in Data-Driven Decision-Making. Available at SSRN 5128174.

Fatema, K., Sinnappan, P., Meng, C.S. and Watabe, M. 2024. Technological advancements and innovations in the tourism industry: Driving sustainable tourism. *In*: The Need for Sustainable Tourism in an Era of Global Climate Change: Pathway to a Greener Future. 121–149. Emerald Publishing Limited.

Fornell, C. and Larcker, D.F. 1981. Evaluating structural equation models with unobservable variables and measurement error. Journal of Marketing Research. 18(1): 39–50.

George, A.S. 2024. Automated Futures: Examining the Promise and Peril of AI on Jobs, Productivity, and Work-Life Balance. Partners Universal Innovative Research Publication. 2(6): 1–17. https://doi.org/10.5281/zenodo.14544519

Gretzel, U. 2021. Conceptualizing the smart tourism mindset: Fostering utopian thinking in smart tourism development. Journal of Smart Tourism. 1(1): 3–8.

Gretzel, U. and Koo, C. 2021. Smart tourism cities: A duality of place where technology supports the convergence of touristic and residential experiences. Asia Pacific Journal of Tourism Research. 26(4): 352–364.

Gretzel, U., Werthner, H., Koo, C. and Lamsfus, C. 2015. Conceptual foundations for understanding smart tourism ecosystems. Computers in Human Behavior. 50: 558–563.

Gün, S. 2025. Artificial Intelligences Impact on Branding in Tourism. *In*: Strategic Brand Management in the Age of AI and Disruption. 155–202. IGI Global Scientific Publishing.

Gupta, P. and Pareek, R. 2024. AI-Enabled Guest Experience: Disrupting Hospitality Through Next-Generation AI Technologies. *In*: Technology and Luxury Hospitality. 165–181. Routledge.

Gursoy, D. 2025. Artificial intelligence (AI) technology, its applications and the use of AI powered devices in hospitality service experience creation and delivery. International Journal of Hospitality Management. 104212. https://doi.org/10.1016/j.ijhm.2025.104212

Ha, M.T. 2021. The impact of customer experience on customer satisfaction and customer loyalty. Turkish Journal of Computer and Mathematics Education. 12(14): 1027–1038.

Hair, J.F., Risher, J.J., Sarstedt, M. and Ringle, C.M. 2019. When to use and how to report the results of PLS-SEM. European Business Review. 31(1): 2–24.

Hanji, S., Navalgund, N., Hanji, S., Goud, V., Hatture, S. and Wadawadagi, R. 2024. Rise of the machines: Investigating service robot adoption intentions in hotel industry. *In*: AIP Conference Proceedings. 3131: 1. AIP Publishing. https://doi.org/10.1063/5.0229719

Hao, X. and Demir, E. 2024. Artificial intelligence in supply chain decision-making: an environmental, social, and governance triggering and technological inhibiting protocol. Journal of Modelling in Management. 19(2): 605–629.

Harun, M.K., Mokhtar, A.Z., Singh, R.K.S. and Rahman, N.A.A. 2022. Artificial intelligence, biometric technology, and mobile technology application in aviation and tourism: How it changes the landscape of tourism and air travel. *In*: Digital Transformation in Aviation, Tourism and Hospitality in Southeast Asia. 63–71. Routledge.

Henseler, J., Ringle, C.M. and Sinkovics, R.R. 2009. The use of partial least squares path modeling in international marketing. *In*: Sinkovics, R.R. and Ghauri, P.N. (eds). New Challenges to International Marketing. 277–319. Emerald Group Publishing Limited. https://doi.org/10.1108/S1474-7979(2009)0000020014

Huang, M.H. and Rust, R.T. 2018. Artificial intelligence in service. Journal of Service Research. 21(2): 155–172.

Huang, M.H. and Rust, R.T. 2021. A strategic framework for artificial intelligence in marketing. Journal of the Academy of Marketing Science. 49(1): 30–50.

Huddin, M.N., Kurnia, D., Deviyantoro, D. and Nafiudin, N. 2024. The relationship between customer experience, customer satisfaction, customer trust, and customer loyalty in tourism destination. Diponegoro International Journal of Business. 6(2): 142–154. https://doi.org/10.14710/dijb.6.2.2023.142-154

Ifekanandu, C.C., Anene, J.N., Iloka, C.B. and Ewuzie, C.O. 2023. Influence of artificial intelligence (AI) on customer experience and loyalty: Mediating role of personalisation. Journal of Data Acquisition and Processing. 38(3): 1936.

Inavolu, S.M. 2024. Exploring AI-driven customer service: Evolution, architectures, opportunities, challenges and future directions. International Journal of Engineering and Advanced Technology. 13(3): 156–163.

Ioannou, I., Kassinis, G. and Papagiannakis, G. 2023. The impact of perceived greenwashing on customer satisfaction and the contingent role of capability reputation. Journal of Business Ethics. 185(2): 333–347. https://doi.org/10.1007/s10551-022-05151-9

Ioku, T., Song, J. and Watamura, E. 2024. Trade-offs in AI assistant choice: Do consumers prioritize transparency and sustainability over AI assistant performance? Big Data and Society. 11(4): 20539517241290217.

Ionescu, A.M. and Sârbu, F.A. 2024. Exploring the impact of smart technologies on the tourism industry. Sustainability. 16(8): 3318. https://doi.org/10.3390/su16083318

Ivanov, S. and Webster, C. 2019. Perceived appropriateness and intention to use service robots in tourism. *In*: Pesonen, J. and Neidhardt, J. (eds). Information and Communication Technologies in Tourism 2019. Cham: Springer. https://doi.org/10.1007/978-3-030-05940-8_19.

Ivanov, S., Gretzel, U., Berezina, K., Sigala, M. and Webster, C. 2019. Progress on robotics in hospitality and tourism: A review of the literature. Journal of Hospitality and Tourism Technology. 10(4): 489–521. https://doi.org/10.1108/JHTT-08-2018-0087

Jeong, M. and Shin, H.H. 2020. Tourists' experiences with smart tourism technology at smart destinations and their behavior intentions. Journal of Travel Research. 59(8): 1464–1477. https://doi.org/10.1177/0047287519883034

Kaakandikar, R.M., Patel, Y. and Tripathy, L. 2025. Sustainability and Ethical Tourism in the Digital Age: Harnessing AI and the Metaverse. *In*: Redefining Tourism With AI and the Metaverse. 123–162. IGI Global Scientific Publishing.

Kabadayi, S., Ali, F., Choi, H., Joosten, H. and Lu, C. 2019. Smart service experience in hospitality and tourism services: A conceptualization and future research agenda. Journal of Service Management. 30(3): 326–348. https://doi.org/10.1108/JOSM-11-2018–0377

Kim, J.H., Kim, J., Kim, S. and Hailu, T.B. 2024. Effects of AI ChatGPT on travelers' travel decision-making. Tourism Review. 79(5): 1038–1057. https://doi.org/10.1108/TR-07-2023-0489

Knani, M., Echchakoui, S. and Ladhari, R. 2022. Artificial intelligence in tourism and hospitality: Bibliometric analysis and research agenda. International Journal of Hospitality Management. 107: 103317. https://doi.org/10.1016/j.ijhm.2022.103317

Koo, I., Zaman, U., Ha, H. and Nawaz, S. 2025. Assessing the interplay of trust dynamics, personalisation, ethical AI practices, and tourist behavior in the adoption of AI-driven smart tourism technologies. Journal of Open Innovation: Technology, Market, and Complexity. 11(1): 100455.

Li, L., Yuan, L. and Tian, J. 2023. Influence of online E-commerce interaction on consumer satisfaction based on big data algorithm. Heliyon. 9(8): e18322. https://doi.org/10.1016/j.heliyon.2023.e18322

Liebherr, M., Enkel, E., Law, E., Mousavi, M., Sammartino, M. and Sieberg, P. 2025. Dynamic calibration of trust and trustworthiness in AI-enabled systems. International Journal on Software Tools for Technology Transfer. https://kclpure.kcl.ac.uk/portal/en/publications/dynamic-calibration-of-trust-and-trustworthiness-in-ai-enabled-sy

Lowry, P.B. and Gaskin, J. 2014. Partial least squares (PLS) structural equation modeling (SEM) for building and testing behavioral causal theory: When to choose it and how to use it. IEEE Transactions on Professional Communication. 57(2): 123–146. 10.1109/TPC.2014.2312452

Majid, G.M., Tussyadiah, I., Kim, Y.R. and Pal, A. 2023. Intelligent automation for sustainable tourism: A systematic review. Journal of Sustainable Tourism. 31(11): 2421–2440.

Malkomsen, E.R. 2024. In What Ways Do Different AI Interfaces Affect Perceived Creepiness? Comparing a Social Robot with a Chatbot (Master's thesis). Retrieved from: https://www.duo.uio.no/handle/10852/112603

Manzoor, S.R., Ullah, R., Khattak, A., Ullah, M. and Han, H. 2024. Exploring tourist perceptions of artificial intelligence devices in the hotel industry: Impact of industry 4.0. Journal of Travel and Tourism Marketing. 41(2): 272–291. https://doi.org/10.1080/10548408.2024.2310169

Mascarenhas, O.A., Kesavan, R. and Bernacchi, M. 2006. Lasting customer loyalty: A total customer experience approach. Journal of Consumer Marketing. 23(7): 397–405. https://doi.org/10.1108/07363760610712939

Miceli, A., Hagen, B., Riccardi, M.P., Sotti, F. and Settembre-Blundo, D. 2021. Thriving, not just surviving in changing times: How sustainability, agility and digitalization intertwine with organizational resilience. Sustainability. 13(4): 2052.

Mishra, D., Das, S. and Patnaik, R. 2024. Application of AI technology for the development of destination tourism towards an intelligent information system. Economic Affairs. 69(2): 1083–1095.

Mogaji, E. and Jain, V. 2024. How generative AI is (will) change consumer behaviour: Postulating the potential impact and implications for research, practice, and policy. Journal of Consumer Behaviour. 23(5): 2379–2389. https://doi.org/10.1002/cb.2345

Moghaddasnian, A.H. 2024. Artificial intelligence and the great transformation in the tourism industry revolutionizing travel services. Journal of Business Data Science Research. 3(01): 36–47.

Nadeem, M. 2025. From algorithms to authenticity: Ensuring ethical customer engagement in the age of artificial intelligence. International Journal of Business and Management. 20(2).

Nama, N.P. 2021. Enhancing user experience in mobile applications through AI-driven personalisation and adaptive learning algorithms. World Journal of Advanced Engineering Technology and Sciences. 3(2): 083–094.

Nguyen, V.T., Phong, L.T. and Chi, N.T.K. 2023. The impact of AI chatbots on customer trust: an empirical investigation in the hotel industry. Consumer Behavior in Tourism and Hospitality. 18(3): 293–305. https://doi.org/10.1108/CBTH-06-2022-0131

Nunnally, J.C. 1978. An overview of psychological measurement. Clinical Diagnosis of Mental Disorders: A Handbook. 97–146.

Nwanna, M., Offiong, E., Ogidan, T., Fagbohun, O., Ifaturoti, A. and Fasogbon, O. 2025. AI-driven personalisation: Transforming user experience across mobile applications. Journal of Artificial Intelligence, Machine Learning and Data Science. 3(1): 1930–1937.

Oprea, S.V., Nica, I., Bâra, A. and Georgescu, I.A. 2024. Are skepticism and moderation dominating attitudes toward AI-based technologies? American Journal of Economics and Sociology. 83(3): 567–607.

Patil, D. 2024. Artificial intelligence-driven customer service: Enhancing personalization, loyalty, and customer satisfaction. Loyalty, And Customer Satisfaction. http://dx.doi.org/10.2139/ssrn.5057432

Patrichi, I.C. 2024. AI solutions for sustainable tourism management: a comprehensive review. Journal of Information Systems and Operations Management. 18(1): 172–185.

Pencarelli, T. 2020. The digital revolution in the travel and tourism industry. Information Technology and Tourism. 22(3): 455–476.

Pillai, R. and Sivathanu, B. 2020. Adoption of AI-based chatbots for hospitality and tourism. International Journal of Contemporary Hospitality Management. 32(10): 3199–3226. https://doi.org/10.1108/IJCHM-04-2020-0259

Polo-Peña, A.I., Frías-Jamilena, D.M., Peco-Torres, F. and Rodríguez-Molina, M.Á. 2025. Experience and artificial intelligence in hospitality and tourism: a review of reviews and a bibliometric analysis. International Journal of Contemporary Hospitality Management. https://doi.org/10.1108/IJCHM-07-2024-0988

Rajaobelina, L., Tep, S.P., Arcand, M. and Ricard, L. 2021. Creepiness: Its antecedents and impact on loyalty when interacting with a chatbot. Psychology & Marketing. 38(12): 2339–2356. https://doi.org/10.1002/mar.21548

Rane, N. 2023. Enhancing customer loyalty through Artificial Intelligence (AI), Internet of Things (IoT), and Big Data technologies: improving customer satisfaction, engagement, relationship, and experience. Internet of Things (IoT), and Big Data Technologies: Improving Customer Satisfaction, Engagement, Relationship, and Experience. http://dx.doi.org/10.2139/ssrn.4616051

Rane, N., Choudhary, S. and Rane, J. 2023. Sustainable tourism development using leading-edge artificial intelligence (AI), blockchain, internet of things (IoT), augmented reality (AR) and virtual reality (VR) technologies. Blockchain, Internet of Things (IoT), Augmented Reality (AR) and Virtual Reality (VR) Technologies. http://dx.doi.org/10.2139/ssrn.4642605

Rane, N., Choudhary, S.P. and Rane, J. 2024a. Acceptance of artificial intelligence: key factors, challenges, and implementation strategies. Journal of Applied Artificial Intelligence. 5(2): 50–70. https://doi.org/10.48185/jaai.v5i2.1017

Rane, N., Choudhary, S. and Rane, J. 2024b. Artificial intelligence, machine learning, and deep learning for sentiment analysis in business to enhance customer experience, loyalty, and satisfaction. Available at SSRN 4846145.

Rane, N., Paramesha, M., Choudhary, S. and Rane, J. 2024c. Artificial intelligence in sales and marketing: Enhancing customer satisfaction, experience and loyalty. Experience and Loyalty. http://dx.doi.org/10.2139/ssrn.4831903

Rani, Y.A., Balaram, A., Sirisha, M.R., Nabi, S.A., Renuka, P. and Kiran, A. 2024. AI Enhanced Customer Service Chatbot. *In*: 2024 International Conference on Science Technology Engineering and Management (ICSTEM). 1–5. IEEE.

Sağbaş, M. and Aydogan, S. 2025. Unveiling the nuances: How fuzzy set analysis illuminates passenger preferences for AI and human agents in airline customer service. Tourism and Hospitality. 6(1): 43. https://doi.org/10.3390/tourhosp6010043

Samala, N., Katkam, B.S., Bellamkonda, R.S. and Rodriguez, R.V. 2022. Impact of AI and robotics in the tourism sector: A critical insight. Journal of Tourism Futures. 8(1): 73–87. https://doi.org/10.1108/JTF-07-2019-0065

Santos, C., Coelho, A. and Marques, A. 2024. The greenwashing effects on corporate reputation and brand hate, through environmental performance and green perceived risk. Asia-Pacific Journal of Business Administration. 16(3): 655–676. https://doi.org/10.1108/APJBA-05-2022-0216

Sardesai, S., D'Souza, E. and Govekar, S. 2024. Analysing the impacts of artificial intelligence service quality and human service quality on customer satisfaction and customer loyalty in the hospitality sector. Turizam. 28(1): 37–48.

Saunders, M.N. and Darabi, F. 2024. Using multi-and mixed methods research designs. *In*: Parker, J., Donnelly, N., Ressia, S. and Gavin, M. (eds). Field Guide to Researching Employment and Industrial Relations. 71–87. Edward Elgar Publishing.

Saydam, M.B., Arici, H.E. and Koseoglu, M.A. 2022. How does the tourism and hospitality industry use artificial intelligence? A review of empirical studies and future research agenda. Journal of Hospitality Marketing and Management. 31(8): 908–936. https://doi.org/10.1080/19368623.2022.2118923

Semwal, R., Tripathi, N., Rana, A., Dafouti, B.S., Bairwa, M.K. and Mathur, V. 2024. AI-powered personalization and emotional intelligence integration for enhanced service marketing in transformative tourism experiences. *In*: 2024 International Conference on Communication, Computer Sciences and Engineering (IC3SE). 1851–1856. IEEE.

Shahrizoda, S. and Nargiza, A. 2024. The influence of modern technology and artificial intelligence in tourism industry. Science and Innovation. 3(Special Issue 28): 640–644.

Sharma, V. 2024. AI in hospitality and tourism promotion: Opportunities and challenges. Impact of AI and Tech-Driven Solutions in Hospitality and Tourism. 43–65.

Sharmin, F., Sultan, M.T., Badulescu, A., Bac, D.P. and Li, B. 2020. Millennial tourists' environmentally sustainable behavior towards a natural protected area: An integrative framework. Sustainability. 12(20): 8545.

Shi, S., Gong, Y. and Gursoy, D. 2021. Antecedents of trust and adoption intention toward artificially intelligent recommendation systems in travel planning: A heuristic–systematic model. Journal of Travel Research. 60(8): 1714–1734. https://doi.org/10.1177/0047287520966395

Singh, P. and Singh, V. 2024. The power of AI: enhancing customer loyalty through satisfaction and efficiency. Cogent Business and Management. 11(1): 2326107.

Solakis, K., Katsoni, V., Mahmoud, A.B. and Grigoriou, N. 2024. Factors affecting value co-creation through artificial intelligence in tourism: a general literature review. Journal of Tourism Futures. 10(1): 116–130. https://doi.org/10.1108/JTF-06-2021-0157

Song, Y. and He, Y. 2023. Toward an intelligent tourism recommendation system based on artificial intelligence and IoT using Apriori algorithm. Soft Computing. 27(24): 19159–19177. https://doi.org/10.1007/s00500-023-09330-2

Tenakwah, E.S. and Watson, C. 2025. Embracing the AI/automation age: preparing your workforce for humans and machines working together. Strategy and Leadership. 53(1): 32–48. https://doi.org/10.1108/SL-05-2024-0040

Topsakal, Y. 2025. Artificial intelligence-based sustainable tourism planning: A conceptual model proposal. *In*: Advancing Smart Tourism Through Analytics. 65–94. IGI Global Scientific Publishing.

Tuo, Y., Wu, J., Zhao, J. and Si, X. 2025. Artificial intelligence in tourism: Insights and future research agenda. Tourism Review. 80(4): 793–812. https://doi.org/10.1108/TR-03-2024-0180

Tuomi, A., Tussyadiah, I.P. and Stienmetz, J. 2021. Applications and implications of service robots in hospitality. Cornell Hospitality Quarterly. 62(2): 232–247. https://doi.org/10.1177/1938965520923961

Tussyadiah, I. 2020. A review of research into automation in tourism: Launching the annals of tourism research curated collection on Artificial Intelligence and robotics in tourism. Annals of Tourism Research. 81: 102883.

Vashishth, T.K., Sharma, V., Sharma, K.K., Kumar, B., Kumar, A. and Panwar, R. 2024. Artificial intelligence (AI)–powered chatbots: Providing instant support and personalized recommendations to guests 24/7. *In*: Technology and Luxury Hospitality. 211–236. Routledge.

Vashishth, T.K., Sharma, K.K., Kumar, B., Chaudhary, S. and Panwar, R. 2025a. Enhancing customer experience through AI-enabled content personalization in e-commerce marketing. Advances in Digital Marketing in the Era of Artificial Intelligence. 7–32.

Vashishth, T.K., Sharma, V., Sharma, M.K. and Sharma, R. 2025b. Enhancing hotel customer service with AI-powered chatbots. *In*: Marketing Technology-Infused Hospitality: Upskilling Frontline Employees for Competitiveness. 83–114. IGI Global Scientific Publishing.

Venkateswaran, P.S., Dominic, M.L., Agarwal, S., Oberai, H., Anand, I. and Rajest, S.S. 2024. The role of artificial intelligence (AI) in enhancing marketing and customer loyalty. *In*: Singh, S., Rajest, S., Hadoussa, S., Obaid, A. and Regin, R. (eds). Data-Driven Intelligent Business Sustainability. 32–47. IGI Global Scientific Publishing.

Waladi, C., Lamarti, M.S. and Khaldi, M. 2024. Transforming Customer Experiences with Customer-Centric AI-Powered Chatbots in the Fourth Industrial Revolution. *In*: 2024 Mediterranean Smart Cities Conference (MSCC). 1–5. IEEE.

Wijethilak, H.P., Singh, S.K. and Dhand, S. 2025. AI for personalizing customer interactions: Improving engagement and satisfaction. *In*: Tunio, M. (ed.). AI, Corporate Social Responsibility, and Marketing in Modern Organizations. 221–244. IGI Global Scientific Publishing.

Wong, L.W., Tan, G.W.H., Ooi, K.B. and Dwivedi, Y. 2024. The role of institutional and self in the formation of trust in artificial intelligence technologies. Internet Research. 34(2): 343–370. https://doi.org/10.1108/INTR-07-2021-0446

Xia, Z. and Shannon, R. 2025. Navigating the digital frontier: Exploring the dynamics of customer–brand relationships through AI chatbots. Sustainability. 17(5): 2173. https://doi.org/10.3390/su17052173

Zerilli, J., Bhatt, U. and Weller, A. 2022. How transparency modulates trust in artificial intelligence. Patterns. 3(4). Accessed from: https://www.cell.com/patterns/fulltext/S2666-3899(22)00028-9

Zhou, S., Yi, N., Rasiah, R., Zhao, H. and Mo, Z. 2024. An empirical study on the dark side of service employees' AI awareness: Behavioral responses, emotional mechanisms, and mitigating factors. Journal of Retailing and Consumer Services. 79: 103869. https://doi.org/10.1016/j.jretconser.2024.103869

Chapter 7

AI in Tourism: Exploring the Impact of Technology Readiness Across Pre-, During-, and Post-Travel Phases of Tourists

Bindu Ranga*[1], Indu Ranga[2],
Amal Raneem K[3] and Nagarjuna G[4]

[1]Assistant Professor, FHMCT, Parul University, Vadodara, India
ORCID: 0000-0002-2834-7554

[2]Assistant Professor, Christ (Deemed to be University), Bengaluru, India
ORCID:0000-0002-8946-050X

[3]Department of Commerce and Management Studies
PSMO College, Tirurangadi, Malappuram, Kerala
ORCID:0000-0002-9312-9174

[4]Assistant Professor, School of Business and Management
Christ (Deemed to be University), Bengaluru, India

INTRODUCTION

The use of Artificial Intelligence (AI) in tourism has recently proliferated, as it provides enhanced efficiencies and reduces costs (Hwang et al. 2022, Lv et al. 2022). There has been substantial discourse on tourists' willingness to use AI-enabled innovations—specifically, service robots, autonomous vehicles,

*For Correspondence: bindutourism16@gmail.com

DOI: 10.1201/9781003540083-7

with them more often (Walczuch et al. 2007). Furthermore, the individual's innovativeness results in engagement in technology, which can positively influence the travel experience holistically (Kang and Gretzel 2012). From a travel point of view, individuals with an innovative orientation are generally early adopters of travel related technologies and feel significantly more comfortable experimenting with new technologies or services in travel (Chen et al. 2021). Overall, innovation orientation positively influences online purchase intention (Chen et al. 2021).

Optimism

While innovativeness indicates a person's willingness to adopt technology early, optimism refers to a favorable attitude toward technology's ability to increase efficiency and control (Parasuraman and Colby 2001). Technological optimism is the favorable attitude toward technology where people anticipate it will provide greater control, flexibility, and efficiency in their lives (Parasuraman and Colby 2015). It is a belief that technology improves control, flexibility, and efficiency, and strongly influences adoption behavior (Parasuraman and Colby 2015). Optimists are more likely to try new technology and disregard potential drawbacks (Lu et al. 2012). For instance, in the financial services sector, optimists are more likely to seek alternative investment options, like robo-advisors (Clark-Murphy and Soutar 2004). Optimism relates to the customer's perceived usefulness and ease of use of the technology, as well as higher overall satisfaction (Thong et al. 2006). Such travelers typically value technology-enhanced services more, trust new technology, acknowledge functionality in a positive manner, and are generally more satisfied (Lu et al. 2012, Walczuch et al. 2007).

Discomfort

Technological discomfort occurs when individuals feel overwhelmed or lack control over technology, which can result in their rejection of new systems (Parasuraman and Colby 2015, Tsang et al. 2004). It is termed as a perceived lack of control over technology, leaving the user feeling overwhelmed (Parasuraman 2000). Travelers higher in discomfort tend to find technology-enhanced services (TESs) complex and will be less likely to interact with these in a positive manner (Wang et al. 2017). Their dissatisfaction is the result of the anxiety and uncertainty when utilizing new technologies (Lu et al. 2012). This discomfort can have a deleterious effect on their satisfaction, in addition to their online purchase intention, which is exacerbated by the possibility of discomfort leading to avoiding engaging with technology altogether (Parasuraman 2000).

Insecurity

Insecurity is an outcome of privacy apprehensions, a lack of trust in technology, and discomfort that is driven by an excessive need for control (Pradhan et al. 2018). Technology insecurity is characterized by a lack of trust or skepticism regarding the operation and risks of technology, reducing the willingness to adopt technology (Kaplan and Haenlein 2019, Lu et al. 2012), or increasing the likelihood of nonconformity (Kaplan and Haenlein 2019, Oliveira et al. 2016). These elements all contribute to the consumers' intention to use technology-based services. Insecurity was defined as a lack of trust in technology, with users feeling skeptical about the effectiveness of that technology (Parasuraman 2000). Insecure travelers had higher anxiety and were less likely to trust or use new technologies evaluated by travelers, leading to lower satisfaction levels (Lu et al. 2012). Insecure travelers also tend to overestimate the complications of TESs and have difficulty understanding and using those systems (Wang et al. 2017).

AI-POWERED TOURIST ASSISTANCE

AI-Driven Planning in Pre, Real, and Post-Time Travel Experiences

Tourist behavior is essential to understand for tourism managers who want to attract tourists. Pearce (2005) distinguishes between five stages of tourist behavior: anticipation (pre-purchase); travel to the destination; on-site experience; return travel; and extended recall and recollection. Among these, the pre-purchase stage, essentially the determination of where to go, is particularly noteworthy (Das et al. 2024). Additionally, there are several factors involved in destination selection, including internal factors (such as beliefs, motives, and attitudes) and external factors (such as time, destination characteristics, and perceived costs) that may affect selection (Sonmez and Sirakaya 2002). Destination image, in addition to only the pre-travel phase, may influence travel behavior; a tourist's behavior during and after their visit is also relevant (Tavitiyaman and Qu 2013, Chen et al. 2013). Research conducted by many contemporary scholars showed how destination image positively impacts trip quality, satisfaction, intent to return, and intent to recommend (Batabyal et al. 2023, Bigné et al. 2001). Similarly, Chen and Tsai (2007) found that destination image strongly impacts trip quality and behavioral intentions, such as intent to return and intent to recommend. Boniface and Cooper (2009) further discuss tourism experiences from three different stages, each with their own characteristics. The first stage, anticipation, happens before travel and is shaped by the mental image or concept of the destination. Anticipation arguably has an important role in destination choice. However, incentive travelers typically exert little control over the incentive destination; the sponsor company or incentive planner makes the destination choice. The second stage, realization, characterizes the intention of the trip and includes tourism experiences at the destination. The realization stage also includes impressions developed during the travel process to and from the destination and the overall travel experience (Batabyal et al. 2023, Boniface and Cooper 2009).

The last stage, recollection, occurs after the trip and is what this study is primarily focused on. Recollection is the phase when travelers compare their trip experiences to their travel expectations formed in the anticipation phase. Travelers either report satisfaction and possibly greater intentions to return to this same travel destination, or their travel experiences do not meet or exceed their expectations and result in dissatisfaction. Akhoondnejad (2015) aimed to examine foreign tourists' perceptions of Iran and noted a statistically different perception of Iran from before their trip and after. Tourists reported a negative image of Iran before their trip, then reported a positive perception of Iran after their trip. The post-travel image type of the destination impacted the trip value and tourist satisfaction, which in turn influenced whether or not tourists would go to that destination again or recommend the destination. For travelers, the anticipation stage is started by the announcement of the incentive, which creates interest in the travel destination, while the experience on the ground and post-travel evaluation are equally important because both stages of the travel process directly influence the likelihood of visiting the destination again (Potgieter et al. 2016). As Li et al. (2008) propose, decisions about returning to a destination are not formed merely based on experiences in the destination, but also on thoughts and evaluations of the trip post-travel, which document and reflect on our memory.

TRI and Tourism Experiences

Within tourism contexts, TRI has been influential in their proposed model of consumers' online purchase intentions and e-loyalty towards online travel sites (Shariffuddin et al. 2023). Pradhan et al. (2018) used the Technology Readiness Index (TRI) model to look at how travelers' perceptions of risks and benefits predict their intention to use smart devices while traveling in tourism and indicated the perceived benefits led to higher intention to use smart devices while traveling, and that perceived risks are different based on traveler characteristics.

Optimism and innovativeness were identified as key positive attributes, whereas insecurity and discomfort were negative attributes with induced implications. Optimistic travelers view technology as a tool to help enhance their travel experience, and they are more likely to purchase travel products/services online and generally trust that features inherent to online travel services will perform as intended, which influences their online purchasing intentions and therefore their online loyalty (Kim et al. 2020). Optimistic consumers are willing to utilize technology when traveling and tend to view technology as trustworthy and functional, which can also influence developer adoption (Lu et al. 2012, Walczuch et al. 2007). Victorino et al. (2009) used a Technology Readiness Index (TRI) to profile hotel customers, demonstrating that TRI can offer new levels of market segmentation by linking technology usage behaviors to demographics. They also reported that younger travelers, with the highest TRI scores, preferred using higher star-rated hotels. In terms of expanding the context for TRI, researchers such as (Chung et al. 2015) assessed TRI in the context of augmented reality (AR) and tourism. However, the majority of published studies on TRI only focus on TRI as a personal attribute and have not explored it in relation to other individual factors.

Traveler's use of modular TRI scores can enhance their understanding of technology acceptance or usage in tourism, as high-TR scores are often more correlated between TES, satisfaction, and behavioral intentions (Wang and Sparks 2014).

High innovative travelers are more likely to test digital tools such as online booking sites, and mobile apps for trip planning (Shariffuddin et al. 2023). Studies indicate a positive relationship between innovativeness and travelers' attitudes toward online services or booking sites, with innovative travelers being more open to use online services and online booking sites (Kim and Forsythe 2008). Discomfort is reflected in the tension or belief the individual feels around technology, derived from a lack of control or feeling overwhelmed (Shariffuddin et al. 2023). Discomfort for travelers may manifest in feeling confused or nervous with the booking process of an online service or mobile app, and travelers with higher insecurity may distrust online services, indicating perceived reliability (Shariffuddin et al. 2023). As a result, these travelers may demonstrate less impulse to engage with online purchasing intentions, affecting not only their purchasing intentions but in the long-term, loyalty to online travel sites (Mandolfo and Lamberti 2021). Rodríguez Bolívar et al. (2025) examined privacy concerns, perceived risks, and trust in smart tourism destinations, and found environmental factors mitigate privacy concerns and that prior privacy violations impacted their trust.

Understanding how individuals respond to new technologies involves the important concept of technology readiness; in fact, there are individual differences in personality that influence that readiness to adopt technology (Flavián et al. 2022). Yi et al. (2019) argue for applications that incorporate important travel resources and connect various stakeholders, e.g. information seekers (users), content providers (algorithms), and local merchants (suppliers), through real-time interactions. Zhang et al. (2024b) discussed how ChatGPT 4.0 is transforming and reshaping travel planning and experiences. Their study found that in the pre-travel phase, ChatGPT assists users by providing destination, activity, and transport recommendations tailored to user personality traits. The use of ChatGPT 4.0's image recognition enables the user to receive additional information about objects during travel (i.e. unfamiliar food or wildlife). In the post-travel phase, it also assists the user in social media posts and writing reviews on location visits. Further, they stated that multimodal AI such as ChatGPT can dramatically enhance personalized travel experiences and plans, and change how travelers share their experiences.

Based on the literature, the following hypothesis are examined in the current study:

H1: Perceived AI Insecurity in Travel (ISEC) influences During- Travel (DT) experience.

H2: Perceived AI Insecurity in Travel (ISEC) influences Post-Travel (PsT) experience.

H3: Perceived AI Insecurity in Travel (ISEC) influences Pre-Travel (PrT) experience.

H4: AI Innovativeness in Travel (INN) influences During- Travel (DT) experience.

H5: AI Innovativeness in Travel (INN) influences Post-Travel (PsT) experience.

H6: AI Innovativeness in Travel (INN) influences Pre-Travel (PrT) experience.

H7: Optimism Toward AI in Travel (OPT) influences During-Trip (DT) experience.

H8: Optimism Toward AI in Travel (OPT) influences Post-Travel (PsT) experience.

H9: Optimism Toward AI in Travel (OPT) influences Pre-Travel (PrT) experience.

H10: AI-Induced Discomfort in Travel (DCOM) influences During- Travel (DT) experience

H11: AI-Induced Discomfort in Travel (DCOM) influences Post-Travel (PsT) experience.

H12: AI-Induced Discomfort in Travel (DCOM) influences Pre-Travel (PrT) experience.

RESEARCH METHODOLOGY

Construct Development and Conceptual Framework

In the study's conceptual framework (Figure 7.1), the researchers analyzed the impact of the use of AI tools by travelers on their travel experience, and measured the following independent variables affecting traveler experiences of AI utilization: Innovativeness (INN), Discomfort (DCOM), Optimism (OPT), Insecurity (ISEC), during various stages of their travel: Pre-Travel (PrT), Post-Travel (PsT), and During-Trip (DT). The conceptual framework illustrates how AI is interactively construed over the phases of travel experience and affects travelers' entire experience and decision-making process related to travel.

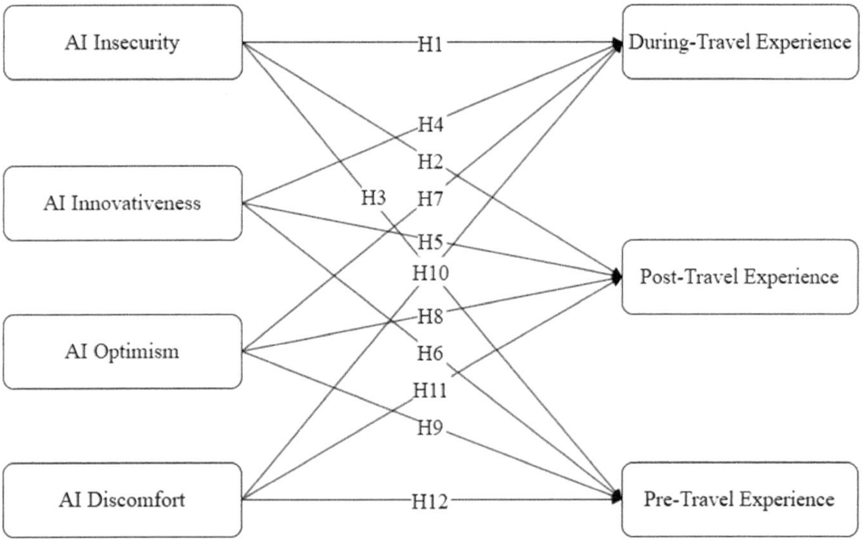

Figure 7.1 Conceptual model (*Source:* Authors' Own Work).

The constructs comprise multiple items, with each item representing some unique aspect of AI domains concerning travel encounters, along with references that grounded the constructs theoretically. Innovativeness (INN) captured the innovative AI travel technology resulting in travel ease and creativity enhancement (Kim et al. 2023, Ribeiro et al. 2022). Discomfort (DCOM) related to the ease of planning using AI and the reduction of stress (Pillai et al. 2020, Shariffuddin et al. 2023). Optimism (OPT) reflected a position of positivity from travelers for AI in enhancing travel experiences (Lu et al. 2012, Parasuraman and Colby 2015). Insecurity (ISEC) was concerned with trusting AI in terms of risk mitigation and privacy (Chi et al. 2021, Seo and Lee 2021). Pre-Travel (PrT) referred to the AI's ability to lessen issues in pre-trip planning and included recommendations and potential points of problem detection (Zhang et al. 2024b, Wang et al. 2023). During-Travel (DT), was focused on real-time AI support for enhancing travel experiences with navigation and language translation (Xu et al. 2024, Yi et al. 2019). Finally, Post-Travel (PsT) assessed AI's role in post-travel experiences, including organizing memories and future trip planning (Wang et al. 2016, Rodriguez-Bolivar et al. 2025).

Data Collection

The constructs were developed, grounded in the literature, and positioned to give a meaningful exploration of AI's implications for travel experiences. In the study, purposive sampling was utilized to survey Gen Z travelers who knowingly used AI and AI-powered technologies during travel. This purposive sampling allowed exploration of a larger net of technologically savvy travelers, increasing the chances of obtaining insightful data for understanding technology influencers on travel experiences. The data were gathered using an online survey through Google Forms, which was highly favorable for gaining access and reaching a broad geographic gap. The survey was distributed through social media and travel forums for digital-origin travelers where travelers actively discuss and share notes on previous travel experiences. The research questionnaire was formulated to elicit demographic information, travel behavior, perspectives of AI-driven tools, and experience-based AI beliefs that were all consistent with the research aims of this study. For analysis purposes, a total of 282 responses were incorporated into the study. Additionally, 17 responses were eliminated due to the presence of erroneously filled-out questionnaires. Partial Least Squares Structural Equation Modeling (PLS-SEM) was employed to analyze the gathered data, as it can assess a combination of multiple predictors and outcomes at once. Further internal consistency reliability testing was done for the data, including Cronbach's alpha and composite reliability (CR).

RESULT ANALYSIS

Measurement and Structural Model

The reliability of individual items was measured through factor loadings (Hair et al. 2019), which required factor loadings above 0.708. Each outer loading in this study was above 0.813, with the highest factor loading at 0.900 (INN3), indicating strong shared variance between each of the items and their associated construct. Accordingly, the items demonstrated that they were measuring what was intended, and to ensure no multicollinearity among items, the researchers evaluated the VIF, which fell below the recommended level of five (Table 7.1).

Table 7.1 Reliability and validity

Construct	Indicator	Original Sample (O)	Sample Mean (M)	Standard Deviation (STDEV)	T Statistics (|O/STDEV|)	P Values	VIF
ISEC	ISEC1	0.886	0.886	0.015	58.559	0	2.987
	ISEC2	0.864	0.864	0.015	55.96	0	2.331
	ISEC3	0.892	0.892	0.015	60.55	0	3.02
	ISEC4	0.859	0.859	0.019	45.415	0	2.319
DT	DT1	0.879	0.879	0.018	49.463	0	2.917
	DT2	0.874	0.874	0.015	57.029	0	2.947
	DT3	0.88	0.88	0.017	51.781	0	2.868
	DT4	0.885	0.885	0.013	67.74	0	2.987
INN	INN1	0.814	0.814	0.02	40.355	0	2.18
	INN2	0.841	0.842	0.016	52.461	0	2.257
	INN3	0.9	0.9	0.012	73.034	0	3.554
	INN4	0.826	0.825	0.02	40.955	0	2.146
	INN5	0.88	0.879	0.017	51.438	0	3.098
OPT	OPT1	0.879	0.88	0.015	58.901	0	2.433
	OPT2	0.834	0.833	0.02	41.839	0	2.097
	OPT3	0.827	0.826	0.034	24.14	0	2.023
	OPT4	0.847	0.847	0.018	46.548	0	2.175
PsT	PsT1	0.857	0.857	0.018	48.853	0	2.336
	PsT2	0.853	0.854	0.016	53.874	0	2.319
	PsT3	0.867	0.867	0.016	53.503	0	2.419
	PsT4	0.843	0.843	0.018	46.163	0	2.238
PrT	PrT1	0.888	0.888	0.014	63.027	0	2.922
	PrT2	0.859	0.859	0.019	45.657	0	2.665
	PrT3	0.892	0.892	0.014	63.586	0	3.043
	PrT4	0.886	0.886	0.015	59.461	0	2.979
DCOM	DCOM1	0.861	0.86	0.019	44.885	0	2.49
	DCOM2	0.813	0.813	0.021	38.022	0	1.841
	DCOM3	0.844	0.844	0.025	34.295	0	2.226
	DCOM4	0.854	0.854	0.021	41.439	0	2.112

Source: Authors' Own Work

To assess construct reliability, both the composite reliability and Cronbach's alpha were calculated for all reflective constructs, and both measures confirmed reliability across all factors. For Discomfort, the Cronbach's alpha was above 0.864; composite reliability was above 0.908. For Insecurity (ISEC), the Cronbach's alpha and composite reliability were 0.898 and 0.929, respectively. For Innovation (INN), the Cronbach's alpha and composite reliability were 0.906 and 0.93, respectively. Table 7.2 shows that the Average Variance Extracted (AVE) for each construct was above the required threshold (between 0.711 and 0.777), thereby sufficiently demonstrating convergent validity.

Table 7.2 Construct reliability and validity

Construct	Cronbach's Alpha	Composite Reliability (rho_a)	Composite Reliability (rho_c)	Average Variance Extracted (AVE)
ISEC	0.898	0.899	0.929	0.766
DT	0.903	0.904	0.932	0.774
INN	0.906	0.907	0.93	0.727
OPT	0.869	0.872	0.91	0.718
PsT	0.878	0.879	0.916	0.731
PrT	0.904	0.908	0.933	0.777
DCOM	0.864	0.866	0.908	0.711

Source: Authors' Own Work

Table 7.3 Fornell-Larcker criterion

	ISEC	DT	INN	OPT	P	PT	DCOM
ISEC	0.875						
DT	0.617	0.88					
INN	0.704	0.612	0.853				
OPT	0.685	0.678	0.693	0.847			
P	0.645	0.752	0.659	0.709	0.855		
PT	0.566	0.71	0.61	0.719	0.683	0.881	
DCOM	0.656	0.607	0.611	0.661	0.687	0.61	0.843

Source: Authors' Own Work

To assess discriminant validity, the researchers used the Heterotrait-Monotrait Ratio (HTMT) criteria, which confirm that each of the constructs in the study model expresses distinct concepts. The HTMT values for the study were all less than 0.85, with the maximum value of 0.845 (During-Trip/Post-Travel), confirming sufficient discriminant validity (Henseler et al. 2015). More importantly, the Fornell-Larcker criteria (Table 7.3) confirmed strong discriminant validity because the square root of the AVE for each of the constructs was greater than the correlations between constructs. Upon confirming the measurement model, the researchers calculated the structural model to test the hypotheses. The R^2 values for the constructs as shown in Table 7.4 (i.e. Post-Travel: 0.616; Pre-Travel: 0.561) were noted to have positive values with good predictive accuracy. The f-square (f^2) values (Table 7.5) indicated meaningful effects, with f^2 values of 0.212 (Optimism for Pre-Travel) and 0.111 (Security for Post-Travel).

As noted earlier, this study used both the Standardized Root Mean Square Residual (SRMR) and Normed Fit Index (NFI) as indicators of model fit (with SRMR values less than 0.08 and NFI values more than 0.9 indicating adequate fit—Hair et al. 2017). Consequently, these findings indicate a model with good reliability and validity, and predictive power, thus offering a reasonable starting point for future models after conducting the structural analysis.

Table 7.4 R-square values

Construct	R-Square	R-Square Adjusted
DT	0.534	0.527
PsT	0.616	0.611
PrT	0.561	0.555

Source: Authors' Own Work

Table 7.5 F-square values

	DT	Pst	PrT
ISEC	0.02	0.011	0
INN	0.02	0.038	0.023
OPT	0.103	0.1	0.212
DCOM	0.035	0.111	0.042

Source: Authors' Own Work

The results (Table 7.6) show that significant relationships exist among the latent variables and state the importance of key dimensions and their influence on travel stages. The path coefficients indicate noteworthy effects. Insecurity (H1, H2, H3) reported limited and non-significant effects on During-Trip (DT) ($\beta = 0.152$, $p = 0.09$), Post-Travel (PsT) ($\beta = 0.105$, $p = 0.172$), and Pre-Travel (PrT) ($\beta = -0.005$, $p = 0.953$), meaning tourists experiencing AI insecurity had no noteworthy effects on each of the three stages of travel mentioned. In contrast, Innovation (H4, H5, H6) reported moderate, positive, and significant effects on DT ($\beta = 0.151$, $p = 0.037$), PsT ($\beta = 0.188$, $p = 0.003$), and PrT ($\beta = 0.158$, $p = 0.006$), indicating that tourists who are ready to adopt new technologies and engage with technology more often have a significant effect on their overall travel experiences at each travel stage. Also, similar to Innovation, Optimism (H7, H8, H9) indicated strong, positive, and significant results for DT ($\beta = 0.346$, $p < 0.001$), PsT ($\beta = 0.309$, $p < 0.001$), and PrT ($\beta = 0.482$, $p < 0.001$), indicating that the presence of an optimistic attitude towards technology in the form of AI affects their travel planning through to the post-trip stage.

Lastly, Discomfort (H10, H11, H12) showed both positive and significant effects for DT ($\beta = 0.186$, $p = 0.007$), PsT ($\beta = 0.299$, $p < 0.001$), and PrT ($\beta = 0.197$, $p = 0.009$), thus outlining discomfort as a salient factor in terms of influencing the travel experience. In summary, all the results indicate that tourists who welcome innovation and are optimistic towards technology and AI have a higher impact on their travel decision-making, while tourists who feel insecure towards technological advancements hardly use AI during holidays, and it has little effect on their decisions and choices related to travel. Although tourists might feel discomfort using AI, it still has a significant effect on their travel decisions.

Table 7.6 Hypothesis results

Hypothesis	Hypothesis Path	Original Sample (O)	Sample Mean (M)	Standard Deviation (STDEV)	T Statistics (|O/STDEV|)	P Values	Confidence-Interval
H1	ISEC → DT	0.152	0.151	0.09	1.695	0.09	−0.025, 0.328
H2	ISEC → Pst	0.105	0.102	0.077	1.368	0.172	−0.049, 0.256
H3	ISEC → PrT	−0.005	−0.004	0.078	0.059	0.953	−0.156, 0.15
H4	INN → DT	0.151	0.153	0.072	2.085	0.037	0.019, 0.301
H5	INN → Pst	0.188	0.19	0.064	2.929	0.003	0.065, 0.321
H6	INN → PrT	0.158	0.159	0.057	2.751	0.006	0.049, 0.275
H7	OPT → DT	0.346	0.346	0.082	4.209	0	0.176, 0.505
H8	OPT → Pst	0.309	0.311	0.074	4.186	0	0.169, 0.458
H9	OPT → PrT	0.482	0.483	0.069	6.974	0	0.343, 0.612
H10	DCOM → DT	0.186	0.187	0.069	2.709	0.007	0.051, 0.32
H11	DCOM → Pst	0.299	0.298	0.07	4.278	0	0.159, 0.439
H12	DCOM → PrT	0.197	0.197	0.075	2.631	0.009	0.047, 0.346

Source: Authors' Own Work

FINDINGS AND DISCUSSION

The study explored the influence of travelers' technology readiness, specifically associated with AI—optimism, innovativeness, discomfort, and insecurity—on their travel experiences in the pre-travel, during-trip, and post-trip phases. The findings reveal that the technology readiness dimensions were used differently to shape AI use across different phases of travel. Optimism (OPT) was the most influential technology readiness dimension and positively impacted AI use in all phases of the trip. The strongest effect was found in pre-travel, indicating that optimistic travelers used AI tools in trip planning, such as navigating virtual assistants or employing dynamic pricing tools for accommodations. While traveling, the impact of optimism was also evident; it is possible that the travelers had developed trust in AI tools for navigation, real-time recommendations, or booking modifications. After traveling, the effect of optimism was ongoing, as travelers engaged with AI tools for providing feedback, curating memories, and even writing reviews. Optimistic travelers tend to value technology services more, develop trust in new technology, view functionality positively, and experience satisfaction (Walczuch et al. 2007, Lu et al. 2012). This implies that if travelers can develop positive perceptions of AI, then they will, in turn, use it more frequently in their travel experience. Individuals with an innovative orientation are generally early adopters of travel-related technologies and feel significantly more comfortable experimenting with new technologies or services in travel (Chen et al. 2021).

Innovativeness (INN) consistently exhibited a favorable effect, notably in the context of post-trip AI application, where technologically savvy travelers may utilize AI to deliver individualized travel summaries or suggest automated sharing features. The effects observed in utilization during pre-travel or during-trip contexts were slightly weaker but still maintained significance and reinforced that innovative travelers were early adopters of AI applications. Travelers with a more innovative orientation are typically early adopters of travel-related technology and are much more comfortable trying new technologies or services in travel (Chen et al. 2021). Based on this, it would seem that tourism providers should engage this segment of travelers and encourage them to actively use advanced AI functionalities to enhance engagement. Discomfort (DCOM) demonstrated a paradoxical effect—discomfort had a negative effect on pre-travel and during-trip travel use, yet a positive influence on travelers' post-trip AI use. This variable implies adoption of discomfort on the part of travelers when tangible post-trip benefits result, such as implied automated refunds or review generation about their previous trip. Discomfort remained a barrier, however, in the earlier stages of the journey context, indicating the importance of building intuitive and transparent AI interfaces to minimize hesitance. Travelers with increased discomfort usually view technology-enhanced services (TESs) as complicated and may have a less favorable interaction (Wang et al. 2017). Lastly, while insecurity (ISEC) had the weakest influence on AI use in general, pre-travel and post-trip AI application appeared to not struggle with insecurity. Only a weak marginal effect was demonstrated during-trip, likely provided by a combination of privacy issues in real-time. The appeal of these findings suggests that fear of security hangs in the balance but does not serve as a key deterrent to AI adoption, provided other factors (i.e. usefulness and ease of use) were key enough.

PRACTICAL AND THEORETICAL IMPLICATIONS OF THE STUDY

Practical Implications

This study has various significant implications for tourism businesses, technology developers, governments, and travelers. Artificial intelligence (AI) is positively permeating the tourism sector; it is already changing service delivery, traveler experiences, and customer satisfaction. The current findings show that AI is changing pre-trip aspects of travel, such as trip planning, booking, and itinerary management. For pre-trip planning, which involves optimism and innovativeness to the greatest degree, it is essential to prioritize AI tools that simplify trip customization for tourists. For tourism businesses, this will require investment in AI operating platforms, and the technology will offer tourists a personalized service. Travel providers, including travel agencies, airlines, and hoteliers, can offer unique recommendations to their customers based on their preferences and previous trips. For example, if AI can track customers' historical browsing behavior, the technology can suggest relevant destinations along with promotional offers. AI through chatbots and virtual assistants can provide 24/7 pre-trip support, making the customer service experience seamless while simultaneously communicating at the customer's convenience. Moreover, flight information, hotel vacancies, and rates could all be integrated into a system to alert travelers to this information in real-time, reducing waiting times in the decision-making process. Similarly, itinerary planning tools could use AI to allow travelers to enter a list of their priorities and then generate an itinerary for them in Google Travel, for example. Lastly, for tourism providers, pre-travel AI processes enable greater customer satisfaction and retention, leading to repeat bookings.

During the trip, optimism is still driving AI adoption; however, discomfort negatively impacts adoption in this phase, which elevates the need for transparency concerning how AI works. The evidence shows that incorporating AI into travel enhances the overall travel experience through real-time assistance to travelers. AI navigational applications, such as Google Maps, have been enhanced with AI applications, including traffic updates and transportation alternatives. AI translators are invaluable to hotels, restaurants, and tour operators to better serve international travelers. Biometric AI devices utilizing facial recognition or fingerprint scanning offer speed and security in travel check-in processes, at airports and airlines, increasing boarding speed and security while reducing lines, congestion, and waiting times. AI-based travel applications can also provide travelers with real-time weather updates as well as information about local events that may affect transportation.

The study reveals that AI provides benefits often beyond the trip. To drive adoption and convert discomfort into positive post-trip experiences, developers and providers should emphasize the usability, convenience, and transformative nature of AI to simplify tasks. For example, travelers can use AI-enabled platforms to organize and share their memories, such as AI features in Google Photos. AI also makes it easier for tourism businesses to obtain feedback. Delivering an automatically set-up survey or review request can be a straightforward way to remind travelers to share their feedback. AI-enabled sentiment analytics tools can analyze reviews to identify service strengths and areas for improvement. In addition, AI-enabled CRM tools allow travel businesses to conduct personalized post-travel marketing campaigns for the traveler.

As artificial intelligence grows in the tourism sector, the question of data security becomes one of concern. The research emphasizes the need for strong data protection. Travel organizers should incorporate encryption protocols and secure payment systems to shield sensitive data. AI-led fraud detection systems scan for suspicious activity, such as unauthorized transactions, to offer fraud protection for travelers. Providing travelers with control over their data, by allowing them to adjust their privacy settings, goes a long way towards establishing trust and confidence in any AI-led services. The study also details how AI-led predictive analytics enables travel companies to gain insights into customers' preferences and behaviors, thereby enabling marketers to effectively target promotions and advertising. By harnessing the power of AI for personalized marketing approaches, tourism companies will be able to positively affect conversion rates, ultimately leading to higher levels of customer satisfaction.

Theoretical Implications

This study adds to the theoretical depth of the literature by introducing AI factors into travel experience models, including the examination of how travelers use and are influenced by AI for tourism-related decision-making. It also contributes to the Technology Acceptance Model (TAM) by showing that innovativeness and optimism are not the only factors to consider when determining whether to accept AI in tourism; discomfort is also a vital consideration. Travelers having optimistic attitude towards AI systems are more likely to accept AI-enabled travel apps and often find these apps to be acceptable even if they don't fully understand how AI works. However, concerns related to data privacy can influence the AI system's trustworthiness for travel app users. Further, this research enhances the understanding how AI supports travelers' decision-making before, during, and after their trips. Additionally, it provides the service industry with a strong framework by merging traditional service and customer experience models with technology.

CONCLUSION

The study provides valuable insights into how travelers' technology readiness affects the use of AI tools in multiple trip stages. Optimism was found to be the most influential factor, significantly increasing AI use in pre-trip planning, during-trip navigation, and post-trip reviews. Innovativeness appeared to be a crucial factor, particularly in the post-trip stage, where technologically savvy travelers took advantage of AI, suggesting opportunities for service providers to appeal to early adopters in the market with advanced features. The results suggests the need to address discomfort in early trip stages through intuitive and/or user-friendly AI tools. Lastly, insecurity exerted the least influence on AI use, in real-time context, due to concerns regarding privacy; however, the actionable and easy-to-use nature of AI tools suggested that the perceived benefits outweighed security fears when using AI tools.

Together, these results underscore the importance of enabling tourism stakeholders to tailor AI solutions according to travelers' psychological readiness: alleviating anxiety in discomfort-prone travelers by enhancing user-friendly functionality, offering optimistic travelers clear and tangible benefits, and innovating to meet the needs of tech-savvy users. The tourism industry can increase AI adoption rates, improve customer experiences, and encourage consistent engagement throughout the travel experience by customizing adoption pathways for different traveler segments.

LIMITATIONS AND FUTURE RESEARCH

This research also had its limitations. First, this article studied implications of AI for tourism in the current temporal space; however, due to the speed at which technology develops, new trends could emerge in AI applications beyond those listed here. Secondly, the study primarily investigated the effects of AI on traveler experiences and did not address workforce-related issues such as the risk of job loss or the need to up-skill the workforce in the tourism space. Future research ought examine the impacts of AI on tourism employment and the transformation of the workforce. Future inquiry ought to evaluate the consequences of AI on tourism employment and workforce. Furthermore, ethical issues regarding the use of AI technologies require more research exposure, particularly with respect to visitor privacy and algorithmic bias. Finally, sustainability is an important area that needs a deeper understanding of cross-cultural impacts of AI, across demographics, to better understand the global ramifications of tourism experiences using AI.

REFERENCES

Akhoondnejad, A. 2015. Analyzing the pre-travel, on-travel, and post-travel behaviors of Iran's first-time visitors. Journal of Travel and Tourism Marketing. 32(8): 1023–1033.

Ariza-Montes, A., Quan, W., Radic, A., Yu, J. and Han, H. 2023. Human values and travelerbehaviors: Metaverse for conferences and meetings. Journal of Travel and Tourism Marketing. 40(6): 490–511. https://doi.org/10.1080/10548408.2023.2263766

Au, W.C.W. and Tsang, N.K.F. 2022. What makes a destination smart? An intelligence-oriented approach to conceptualizing destination smartness. Journal of Travel and Tourism Marketing. 39(4): 448–464. https://doi.org/10.1080/10548408.2022.2116627

Batabyal, D., Rawal, Y.S., Chatterjee, P. and Goswami, S. 2023. Temple service quality model for future pilgrimages in Indian sub-continent through user-generated contents (UGC) in TripAdvisor. International Journal of Tourism Anthropology. 9(2): 113–135.

Bigné, J.E., Sánchez, M.I. and Sánchez, J. 2001. Tourism image, evaluation variables and after purchase behaviour: Inter-relationship. Tourism Management. 22: 607–616. doi:10.1016/S0261-5177(01)00035-8.

Boniface, B. and Cooper, C. 2009. Worldwide Destinations: The Geography of Travel and Tourism. 5th Ed. Oxford: Butterworth-Heinemann.

Cai, D., Li, H. and Law, R. 2022. Anthropomorphism and OTA chatbot adoption: A mixed methods study. Journal of Travel and Tourism Marketing. 39(2): 228–255. https://doi.org/10.1080/10548408.2022.2061672

Chen, C.-F. and Tsai, D. 2007. How destination image and evaluative factors affect behavioral intentions? Tourism Management. 28: 1115–1122. doi:10.1016/j.tourman.2006.07.007.

Chen, P. -J., Hua, N. and Wang, Y. 2013. Mediating perceived travel constraints: The role of destination image. Journal of Travel and Tourism Marketing. 30: 201–211. doi:10.1080/10548408.2013.774914.

Chen, Y.-C., Shang, R.-A. and Li, M.-J. 2014. The effects of perceived relevance of travel blogs' content on the behavioral intention to visit a tourist destination. Computers in Human Behavior. 30: 787–799. https://doi.org/10.1016/j.chb.2013.05.019

Chen, S.C., Li, S.H., Liu, S.C., Yen, D.C. and Ruangkanjanases, A. 2021. Assessing determinants of continuance intention towards personal cloud services: Extending UTAUT2 with technology readiness. Symmetry. 13(3): 467. https://doi.org/10.3390/sym13030467

Chi, O.H., Jia, S., Li, Y. and Gursoy, D. 2021. Developing a formative scale to measure consumers' trust toward interaction with artificially intelligent (AI) social robots in service delivery. Computers in Human Behavior. 118: 106700. https://doi.org/10.1016/j.chb.2021.106700

Chung, N., Han, H. and Joun, Y. 2015. Tourists' intention to visit a destination: The role of augmented reality (AR) application for a heritage site. Computers in Human Behavior. 50: 588–599. https://doi.org/10.1016/j.chb.2015.02.068

Clark-Murphy, M., and Soutar, G.N. 2004. What individual investors value: Some Australian evidence. Journal of Economic Psychology. 25(4): 539–555.

Das, A., Kondasani, R.K.R., Deb, R. and Batabyal, D. 2024. Leveraging user-generated content (UGC) in pre-travel behavior: A PLS-SEM and necessary conditions analysis (NCA) approach. Tourism and Hospitality Research. 14673584241292254.

Flavián, C., Pérez-Rueda, A., Belanche, D. and Casaló, L.V. 2022. Intention to use analytical artificial intelligence (AI) in services – The effect of technology readiness and awareness. Journal of Service Management. 33(2): 293–320.

Hair, J.F., Hult, G.T.M., Ringle, C.M. and Sarstedt, M. 2017. A Primer on Partial Least Squares Structural Equation Modeling (PLS-SEM), 2nd Ed. Sage Publications.

Hair, J.F., Risher, J.J., Sarstedt, M. and Ringle, C.M. 2019. When to use and how to report the results of PLS-SEM. European Business Review. 31(1): 2–24. doi: https://doi.org/10.1108/EBR-11-2018-0203

Henseler, J., Ringle, C.M. and Sarstedt, M. 2015. A new criterion for assessing discriminant validity in variance-based structural equation modeling. Journal of the Academy of Marketing Science. 43(1): 115–135. https://doi.org/10.1007/s11747-014-0403-8

Hwang, J., Kim, H.M., Joo, K.-H. and Kim, J.J. 2022. The antecedents and consequences of brand authenticity in the restaurant industry: Robot service employees versus human service employees. Journal of Travel and Tourism Marketing. 39(2): 256–270. https://doi.org/10.1080/10548408.2022.2061678

Kang, M. and Gretzel, U. 2012. Perceptions of museum podcast tours: Effects of consumer innovativeness, Internet familiarity, and podcasting affinity on performance expectancies. Tourism Management Perspectives. 4: 155–163.

Kaplan, A. and Haenlein, M. 2019. Siri, Siri, in my hand: Who's the fairest in the land? On the interpretations, illustrations, and implications of artificial intelligence. Business Horizons. 62(1): 15–25.

Kim, J. and Forsythe, S. 2008. Adoption of virtual try-on technology for online apparel shopping. Journal of Interactive Marketing. 22(2): 45–59.

Kim, J.J., Lee, M.J. and Han, H. 2020. Smart hotels and sustainable consumer behavior: Testing the effect of perceived performance, attitude, and technology readiness on word-of-mouth. International Journal of Environmental Research and Public Health. 17(20): 7455. https://doi.org/10.3390/ijerph17207455

Kim, H., Kim, S.-E., Park, K. and Tennessee, S. 2023. Exploring the role of flow experience and telepresence in virtual reality (VR) concerts. Journal of Travel and Tourism Marketing. 40(7): 568–582. https://doi.org/10.1080/10548408.2023.2276437

Kwon, H.S. and Chidambaram, L. 2000. A test of the technology acceptance model: The case of cellular telephone adoption. Proceedings of the 33rd Annual Hawaii International Conference on System Sciences, Maui, HI, USA, 4–7 January 2000. IEEE. https://doi.org/10.1109/HICSS.2000.926607

Lee, W., Castellanos, C. and Choi, H.C. 2012. The effect of technology readiness on customers' attitudes toward self-service technology and its adoption: The empirical study of US airline self-service check-in kiosks. Journal of Travel and Tourism Marketing. 29(8): 731–743. https://doi.org/10.1080/10548408.2012.730934

Leung, Y. 2013. Perceived benefits. *In*: Gellman, M.D. and Turner, J.R. (eds). Encyclopedia of behavioral medicine. 1450–1451. Springer. https://doi.org/10.1007/978-1-4419-1005-9_394

Li, X.R., Cheng, C., Kim, H. and Petrick, J.F. 2008. A systematic comparison of first-time and repeat visitors via a two-phase online survey. Tourism Management. 29(1): 278–293.

Liljander, V., Gillberg, F., Gummerus, J. and Van Riel, A. 2006. Technology readiness and the evaluation and adoption of self-service technologies. Journal of Retailing and Consumer Services. 13(3): 177–191.

Lu, J., Wang, L. and Hayes, L.A. 2012. How do technology readiness, platform functionality, and trust influence C2C user satisfaction? Journal of Electronic Commerce Research. 13(1): 50–69.

Lv, L., Huang, M., Guan, D. and Yang, K. 2022. Apology or gratitude? The effect of communication recovery strategies for service failures of AI devices. Journal of Travel and Tourism Marketing. 39(6): 570–587. https://doi.org/10.1080/10548408.2022.2162659

Martens, M., Roll, O. and Elliott, R. 2017. Testing the technology readiness and acceptance model for mobile payments across Germany and South Africa. International Journal of Innovation and Technology Management. 14(6): 1–19.

Mandolfo, M. and Lamberti, L. 2021. Past, present, and future of impulse buying research methods: A systematic literature review. Frontiers in Psychology. 12: 687404.

Oliveira, T., Thomas, M., Baptista, G. and Campos, F. 2016. Mobile payment: Understanding the determinants of customer adoption and intention to recommend the technology. Computers in Human Behavior. 61: 404–414.

Parasuraman, A. 2000. Technology Readiness Index (TRI) a multiple-item scale to measure readiness to embrace new technologies. Journal of Service Research. 2: 307–320. https://doi.org/10.1177/109467050024001

Parasuraman, A. and Colby, C. 2001. Techno-ready Marketing: How and Why Consumers Adopt Technology. Free Press.

Parasuraman, A. and Colby, C.L. 2015. An updated and streamlined technology readiness index: TRI 2.0. Journal of Service Research. 18(1): 59–74.

Pearce, P.L. 2005. Tourist Behaviour: Themes and Conceptual Schemes. Clevedon: Channel View Publications.

Pillai, R., Sivathanu, B. and Dwivedi, Y.K. 2020. Shopping intention at AI-powered automated retail stores (AIPARS). Journal of Retailing and Consumer Services. 57: 102207. https://doi.org/10.1016/j.jretconser.2020.102207

Potgieter, M., Jacobs, M. and Geldenhuys, S. 2016. Incentive travellers' proclivity to revisit an incentive destination. African Journal of Hospitality, Tourism and Leisure, Special Edition. 5(3).

Pradhan, M.K., Oh, J. and Lee, H. 2018. Understanding travelers' behavior for sustainable smart tourism: A technology readiness perspective. Sustainability. 10(11): 4259.

Prodanova, J., San-Martín, S. and Jimenez, N. 2018. Are you technologically prepared for mobile shopping? The Service Industries Journal. 1–23.

Ratchford, B.T. 2020. The history of academic research in marketing and its implications for the future. Spanish Journal of Marketing—ESIC. 24(1): 3–36.

Ribeiro, M.A., Gursoy, D. and Chi, O.H. 2022. Customer acceptance of autonomous vehicles in travel and tourism. Journal of Travel Research. 61(3): 620–636. https://doi.org/10.1177/0047287521993578

Rodríguez Bolívar, M.P., Sepe, F., Nanu, L. and Roberto, F. 2025. Perceptions and challenges of blockchain adoption in the tourism industry: A study on trust, privacy, and security. Journal of Hospitality and Tourism Technology. 16(3): 429–449.

Seo, K.H. and Lee, J.H. 2021. The emergence of service robots at restaurants: Integrating trust, perceived risk, and satisfaction. Sustainability. 13(8): 4431. https://doi.org/10.3390/su13084431

Shariffuddin, N.S.M., Azinuddin, M., Yahya, N.E. and Hanafiah, M.H. 2023. Navigating the tourism digital landscape: The interrelationship of online travel sites' affordances, technology readiness, online purchase intentions, trust, and E-loyalty. Heliyon. 9(8).

Sonmez, S. and Sirakaya, E. 2002. A distorted destination image? The case of Turkey. Journal of Travel Research. 41: 185–196. doi:10.1177/004728702237418.

Tavitiyaman, P. and Qu, H. 2013. Destination image and behavior intention of travelers to Thailand: The moderating effect of perceived risk. Journal of Travel and Tourism Marketing. 30: 169–185. doi:10.1080/10548408.2013.774911.

Tavitiyaman, P., Zhang, X. and Tsang, W.Y. 2022. How tourists perceive the usefulness of technology adoption in hotels: Interaction effect of past experience and education level. Journal of China Tourism Research. 18(1): 64–87. https://doi.org/10.1080/19388160.2020.1801546

Thong, J.Y., Hong, S.-J. and Tam, K.Y. 2006. The effects of post-adoption beliefs on the expectation-confirmation model, for information technology continuance. International Journal of Human-Computer Studies. 64(9): 799–810.

Tsang, M.M., Ho, S.C. and Liang, T.-P. 2004. Consumer attitudes toward mobile advertising: An empirical study. International Journal of Electronic Commerce. 8(3): 65–78.

Victorino, L., Karniouchina, E. and Verma, R. 2009. Exploring the use of the abbreviated technology readiness index for hotel customer segmentation. Cornell Hospitality Quarterly. 50(3): 342–359. https://doi.org/10.1177/1938965509338877

Walczuch, R., Lemmink, J. and Streukens, S. 2007. The effect of service employees' technology readiness on technology acceptance. Information and Management. 44(2): 206–215.

Wang, Y. and Sparks, B. 2014. Technology-enabled services: Importance and role of technology readiness. Tourism Analysis. 19(1): 19–33.

Wang, D., Xiang, Z. and Fesenmaier, D.R. 2016. Smartphone use in everyday life and travel. Journal of Travel Research. 55(1): 52–63. https://doi.org/10.1177/0047287514535847

Wang, Y., So, K.K.F. and Sparks, B.A. 2017. Technology readiness and customer satisfaction with travel technologies: A cross-country investigation. Journal of Travel Research. 56(5): 563–577.

Wang, J., Nguyen, N., Jiang, X., Nguyen, H.V. and Saleem, M.A. 2023. Consumers' perceived value and use intention of cashless payment in the physical distancing context: Evidence from an Asian emerging market. Asia Pacific Journal of Marketing and Logistics. 35(6): 1513–1531. https://doi.org/10.1108/APJML-05-2022-0408

Wolfe, H.B. 2010. The insecurity of mobile phones. Proceedings of the Informing Science and Information Technology Education Conference. Cassino, Italy. 119–131.

Xu, H., Law, R., Lovett, J., Luo, J.M. and Liu, L. 2024. Tourist acceptance of ChatGPT in travel services: the mediating role of parasocial interaction. Journal of Travel and Tourism Marketing. 41(7): 955–972.

Yi, Y.K., del Vas, G.M. and Muñoz, A. 2019. An integral mobile application for pre-travel, on-site and post-travel stages. International Journal of Information Systems and Tourism (IJIST). 4(1): 7–17.

Zhang, J., Chen, Q., Lu, J., Wang, X., Liu, L. and Feng, Y. 2024a. Emotional expression by artificial intelligence chatbots to improve customer satisfaction: Underlying mechanism and boundary conditions. Tourism Management. 100: 104835. https://doi.org/10.1016/j.tourman.2023.104835

Zhang, J., Mills, D.J. and Huang, H.W. 2024b. Enhancing travel planning and experiences with multimodal ChatGPT 4.0. *In*: Proceedings of the 2024 International Conference on Innovation in Artificial Intelligence. 12–19.

Zhu, Y., Zhang, R., Zou, Y. and Jin, D. 2023. Investigating customers' responses to artificial intelligence chatbots in online travel agencies: The moderating role of product familiarity. Journal of Hospitality and Tourism Technology. 14(2): 208–224. https://doi.org/10.1108/JHTT-02-2022-0041

Chapter **8**

Exploring the Role of Virtual Reality (VR) in Shaping the Future of Tourism: A Comprehensive Review

Saloni Bansal*[1] and Amit Kumar[2]
[1]Ph.D. Scholar, Department of Tourism and Hotel Management, Central University of Haryana
[2]Assistant Professor, Department of Tourism and Hotel Management, Central University of Haryana

INTRODUCTION

Information and Communication Technology (ICT) has significantly impacted various industries, particularly the travel and tourism sector (Beck et al. 2019). Virtual reality is one of ICT's most significant segments (Guttentag 2010). VR can be utilised as a useful technique for destination marketing (Rainoldi et al. n.d.). Modern IT can improve networking and the "virtuality" of tourism organizations by supporting their strategic marketing and management initiatives. In today's time, consumers are becoming more knowledgeable and demanding as they are familiar with emerging technologies and require personalized, flexible, and innovative products (Buhalis n.d., Sinha et al. 2023). Low-cost VR devices and diverse VR materials offer affordable and convenient ways for individuals to enjoy virtual city and destination tours (Morrison et al. 2024, Tussyadiah et al. 2017).

*For Correspondence: saloni.bansal070397@gmail.com, ak.amitmathur@gmail.com

DOI: 10.1201/9781003540083-8

VR may encourage sustainable travel by enabling users to virtually experience destinations, lowering carbon footprints, and consequently minimizing global warming with extended scope for enriching experiences (Lundgren and Lööv n.d., Morrison et al. 2024). The COVID-19 pandemic significantly impacted the tourism industry. There were several restrictions on travel, leading to the declining phase of the tourism industry. VR technology proved to be a promising tool for the revival of the industry and acted as an alternative to actual travel. For example, during COVID-19, the Ministry of Tourism, Government of India, launched the "Dekho Apna Desh" program, which provided information about various tourist locations in India virtually through tourism experts.

Virtual Reality

Virtual reality stands for computer-generated environments that emulate locations or circumstances in which users can submerge themselves and engage, generally with the use of head-mounted displays and motion-tracking technological equipment, providing an immersive experience of being in a different place (Guttentag 2010).

The historical development of virtual reality traces back to early developments such as Morton Heilig's Sensorama in 1962, which was among the first endeavors to stimulate many senses—sight, hearing, smell, and touch—to fabricate a compelling virtual experience (Heilig 1962). However, Jaron Lanier first coined the term virtual reality in the 1980s. Lanier created VR gear, including data gloves and head-mounted displays (HMDs), that focused on immersive engagement (Lanier 2001).

Today, VR applications can be seen in various areas, such as gaming, entertainment, education, healthcare, tourism, and hospitality. VR in education enhances experiential learning by enabling students to explore complicated subjects via virtual simulations (Freina and Ott 2015). In healthcare, VR is utilized for physical therapy, psychological treatments, and phobias via exposure therapy (Maples-Keller et al. 2017). In tourism, VR can be utilized to offer virtual tours of destinations, thereby giving immense opportunities to tourism businesses to improve marketing strategies and influence the travel decision process of potential customers (Tussyadiah et al. 2018).

The advancement of VR technology has the potential to transform many industries, including entertainment, healthcare, and tourism. VR can enhance individual experiences and their interaction with each other and digital content in their everyday lives.

Virtual Reality in the Travel Industry

Virtual Reality has the potential to transform the tourism industry by offering immersive virtual tours of destinations. It offers interactive experiences to travelers before making a purchase decision. Travel business professionals can use this tool to increase marketing of the destinations, in turn increasing visitor engagement and maximizing revenue. Virtual reality allows tourists to virtually explore destinations of historic importance, natural landscapes, resorts, museums, and places of spiritual and cultural importance, fostering a psychological connection with the destinations before actually visiting them (Tussyadiah et al. 2018). Destination marketing organizations (DMOs) can use VR as a potential instrument for effective marketing strategies with the creation of realistic environments and 3-D simulations. VR can be seen as a significant instrument that can influence the destination choice and travel motivation of the travelers by allowing them to remotely visualize the attractions and accommodations (Han et al. 2019, Guttentag 2010).

VR offers a promising tool in offering accessible and inclusive tourism for all. Individuals facing physical, economic, or social impediments, such as senior citizens, individuals with impairments, or people facing financial constraints, can access the destinations virtually from their remote locations, facilitating access to global and cultural heritage (Guttentag 2010, Cheong 1995).

VR can be utilized to preserve natural, historical, and cultural resources by providing travelers access to endangered and inaccessible locations, promoting awareness for global heritage (Bekele et al. 2018). In sustainable tourism development, VR can be recognized as a successful tool for mitigating negative environmental impacts. It can help in preserving fragile ecosystems, thereby reducing carbon footprints and serving as a substitute to excessive tourism (Han et al. 2019). VR is revolutionizing events, exhibits, and travel agencies by adding virtual showrooms, immersive storytelling, and gamified travel experiences to boost consumer engagement and presence (Jung and Dieck 2017).

Figure 8.1 represents the market share of major virtual travel companies such as Dubai 360, Matterport, YouVisit, Google Earth VR, Real Tour Vision (RTV), 360 Cities, TourVista and others. Dubai 360, Matterport and YouVisit together achieve a significant market share of more than 10% in the virtual tourism industry by 2023. The global virtual tourism market size was valued at USD 728.6 million in 2023 and is projected to grow at a CAGR of 26.4% between 2024 and 2032.

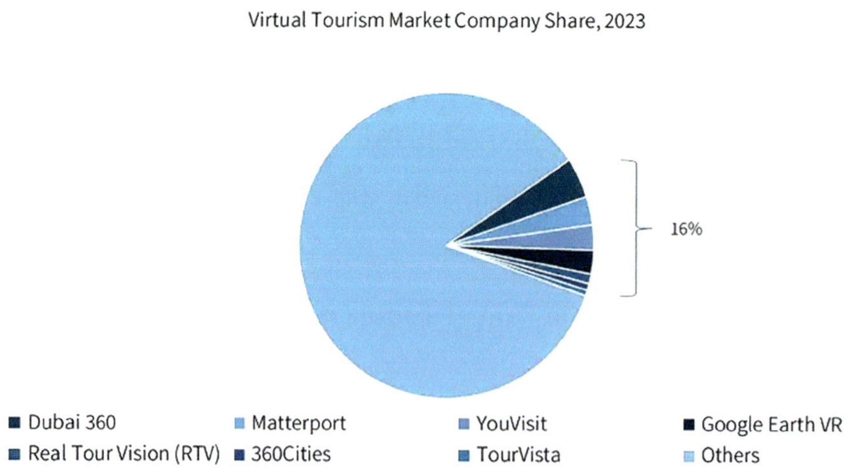

Figure 8.1 Virtual Market Company Share (*Source:* www.gminsights.com.).

LITERATURE REVIEW

(Buhalis et al. 2019) in their paper explored how technological advancements foster value co-creation in the tourism services ecosystem, highlighting the potential disruptive impact of these innovations on market structures.

(Chang and Ku 2023) in their study looked into how tourists' attachment to a place and virtual reality (VR) experiences affected the places they chose to visit. 436 tourists participated in a study that employed a social identity and signaling viewpoint. The interviews were done using purposive sampling. Research showed that while travelers' place attachment has little effect on their cognitive destination image, the VR experience aids in the development of a destination image, which in turn has a notable impact on travelers' choice of destination. By providing passengers with a preview of a destination's appeal, virtual reality can help travel agencies enhance destination marketing. Travel agencies now use virtual reality as a potent marketing tool.

(Florek and Lewicki 2022) in their paper tried to identify the travelers' perspectives about virtual reality during COVID-19 isolation. An online survey was conducted to gather data from Facebook tourist groups who had past experience with VR technology. The findings indicated that VR will likely expand the scope of travel and destination activities and that COVID-19 has influenced people's perceptions and actions toward VR.

(Guttentag 2010) The main applications of VR in tourism were examined in this paper, along with the potential for using VR to offer alternatives to traditional tourism experiences. Various issues and difficulties surrounding VR's integration into tourism were also examined, and several suggestions for further research on VR's applications in tourism were made. The author illustrated the tourism-related uses of VR by summarizing present and upcoming VR technologies. Thus, the author concluded that with VR technologies, the tourism industry is projected to develop in the future. Therefore, it becomes the responsibility of tourism researchers and professionals to take advantage of such opportunities.

(Kostyk and Sheng 2023) in their paper explored the role of VR in customer-centered marketing. The study indicates that VR marketing can effectively aid businesses in achieving their marketing and communication goals.

(Lavuri and Akram 2024) in their study explored the impact of virtual reality on the travel intention of the tourist. Primary data was gathered from 273 Indian samples using purposive sampling. The findings showed that enjoyment, emotional engagement, and a state of flow were all significantly impacted by a genuine virtual reality experience. Likewise, respondents' intention to visit was positively impacted by pleasure, emotional engagement, and a state of flow, with a strong moderating effect of virtual reality attachment on the link between visit intention and pleasure, emotional engagement, and a state of flow.

(Leung et al. 2022) in their study investigated how tourists' memorable VR experiences are influenced by virtual reality (VR) interactive features such as synchronization, two-way communication, and active control, as well as how this influences travelers' intention to continue using VR and their word-of-mouth

(WOM). Survey data from 285 Chinese travelers was collected who had prior virtual reality tourism experience. The findings confirmed that the involvement of VR provides a unique experience, which in turn increases WOM by the tourist. The findings supported the value of creating immersive virtual reality programs, intuitive VR activities, synchronized visitor input with the virtual environment, and two-way visitor-to-virtual environment communication.

(Manoharan et al. 2024) investigated the uses of artificial intelligence (AI) in the travel industry. The findings indicated that AI is predicted to improve traveler experiences, simplify sector administration, give workers realistic tasks, save time, money, and effort, and allow businesses to set competitive prices.

(Mirzaei et al. 2024) investigated the impact of virtual reality (VR) on tourism marketing by examining behavioral intentions and virtual destination image formation. The research was carried out on tourists visiting Iran. For this purpose, a quantitative study was conducted, and a total of 302 participants were surveyed. The results indicated that virtual cognitive and emotive imagery are strongly influenced by the presence in the VR environment, which in turn affects the overall destination image and tourist behavioral intentions.

(Ukpabi et al. 2018) investigated the impact of robotic virtual agents with artificial intelligence (AIRVA) on traveler behavior in the travel and hospitality sector. The research addressed various obstacles to its adoption and provided recommendations for practitioners and academicians to improve its implementation in the travel and hotel industry.

(Oncioiu and Priescu 2022) carried out exploratory research to find out the importance of virtual reality while choosing a destination by the prospective client as well as their expectations, which might help the tourism industry recover from the epidemic. 824 respondents participated in the study. The study revealed that virtual reality applications are utilized for destination marketing to consistently and effectively convey a destination's image to the general audience. The findings also underlined the need to develop future virtual reality technologies for strategic planning in the travel sector.

(Rainoldi et al. 2018) conducted an experimental study to take a look into how virtual reality affects travelers' intentions to visit a place. For this purpose, 101 experiments were carried out to examine in detail the promotional tools used by destinations. According to the research, virtual reality (VR) outperforms traditional promotional materials in terms of speed, interactivity, and detail. The findings can assist destinations in comprehending the value of VR in promoting tourism.

(Ivanov et al. 2017) in their paper examine the acceptance of RAISA by travel agencies and the hospitality industry. They discussed the hardships faced by travel companies while implementing RAISA, including resistance to change and reengineering service processes, and offered recommendations for both companies and robot manufacturers.

(Shukla et al. 2024) The study aimed to investigate the potential of virtual and augmented reality in increasing travel experiences and overcoming physical travel limitations. A survey was conducted on 412 travelers who had utilized AR and VR technologies for travel. Results indicated that VR and AR improve travel by creating immersive experiences. Traditional travel obstacles are removed, yet issues with accessibility, such as cost, still exist. Future acceptance is highly predicted by these technologies' perceived potential, underscoring the necessity of showcasing their immersive advantages.

(Sousa et al. 2024) The study looked into the potential significance of technological innovation on people's opinions of virtual reality travel experiences and, consequently, their propensity to suggest them. The study was conducted at a winery in Portugal's Douro area, which serves as a well-liked tourist destination. Tourists visiting the destination voluntarily participated in the research where they were given VR headsets and were shown a small video. A total of 405 respondents participated in the research. The findings showed that visitors' pleasure and plans for the future are influenced by the quality of a virtual reality encounter. Professionals should thus make investments in VR systems that function properly and create rich material so that visitors will regard the total experience as being of high quality.

(Tussyadiah et al. 2017) investigated the impact of geographical presence in virtual reality settings on perceptions of tourism locations. The goal of the study was to learn more about how VR experiences could influence people's travel choices. The data was collected through an online survey. Factor analysis and ANOVA methods were used for data analysis. The results showed that a considerable contribution to spatial presence came from users' attention allocation to VR settings.

(Wei 2019) examined the key advancements in augmented reality (AR) and virtual reality (VR) research in the travel and hospitality industry and provided directions for future research on AR and VR applications. For this purpose, the author conducted an extensive literature review of AR/VR-related papers published between 2000 and 2018. The study found empirical evidence in the body of existing literature regarding the potential to use VR/AR technology to impact consumer awareness, experience, learning, emotions, visitation, and word-of-mouth intents. This information will help DMOs and marketing firms, especially small ones, decide what they want to achieve with the aid of new technology and incorporate VR into their operational and strategic plans.

(Yersüren and Özel 2024) The study seeks to determine how the quality of virtual reality experiences affects travelers' intentions to visit tourist destinations and take virtual reality trips, using the technology acceptance model (TAM). Quantitative information was obtained from a survey of 198 participants by giving them a hypothetical three-dimensional (3D) travel experience. To collect the data, a total of 198 people were surveyed using a questionnaire. The data was analyzed using the PLS-SEM method. The results showed that perceptions and intentions are influenced by the quality of the virtual reality experience. While perceived risk only influences the intention to travel in virtual reality, perceived usefulness influences attitude and intention to visit the destination, and perceived ease of use influences perceived usefulness. Intentions are influenced by attitude.

(Zlatanov and Popesku 2019) in their paper delve into the utilization of artificial intelligence in the realm of the travel and hospitality industry. They concluded that chatbots and AI-powered travel bots improve tourism offerings, operations, productivity, and product quality.

RESEARCH QUESTIONS

Q1. Does Virtual Reality affect the growth of tourism?
Q2. How can virtual reality help in the future growth of tourism?

Research Objective

To determine the impact of Virtual Reality on the growth of tourism.

Significance

1. Virtual Reality technology has revolutionized the travel industry, allowing travelers to preview destinations, lodging options, and attractions before booking. Hence, VR promotes informed decision-making and lowers uncertainty.
2. Virtual tours offer increased accessibility and are an immersive substitute to traditional travel, particularly during crises such as the COVID-19 pandemic.
3. Travel agencies can use VR as a destination marketing tool, thereby creating opportunities for future growth and development.
4. VR reduces the environmental effect of traditional travel, hence encouraging sustainable travel.

DISCUSSION

Conclusion and Implications

Virtual reality has become a disruptive factor in the travel industry, changing how tour reviews are conducted, destinations are marketed, and issues like accessibility and sustainability are addressed. VR is revolutionizing the travel business, delivering immersive, interactive, and most significantly, tailor-made content that caters to the needs of the latest era of tourists while promoting inclusion and environmental sensitivity. VR's "try-before-you-buy" feature enables travelers to make more informed choices by letting them try out destinations, lodging options, and activities in a virtual location before they commit. That does more than make customers happy, increase traveler confidence, and set companies apart in a crowded market. Using virtual reconstructions and multimedia sagas also does wonders for engagement, meaning it's more than just a boring old stick.

VR really shakes things up when it comes to tours. It breaks down barriers—logistical, financial, and even physical ones—making it possible for people from all walks of life to explore amazing places. It's pretty cool how it opens doors for people with different abilities and backgrounds.

But it's not just about getting people out there; VR also teaches us a thing or two about conservation and responsible travel. It's a neat way to promote sustainable tourism by reducing the environmental footprint that traditional travel often leaves behind.

And let's not forget how the COVID-19 pandemic really put a spotlight on VR. When travel was nearly impossible, virtual experiences helped keep that adventurous spirit alive. Thanks to VR, we could still wander through museums, check out historical sites, and marvel at natural wonders, all from the comfort of our own homes. It's like a ticket to the world, even when the world feels a bit out of reach.

VR can potentially enhance tourism by fostering sustainable behaviors, developing novel stories, and increasing accessibility as it develops. To create a seamless fusion of virtual and physical exploration that meets a wide range of requirements while protecting the environment, the future of tourism rests in making use of VR together with conventional travel. This thorough evaluation offers insightful information for stakeholders, scholars, and travelers alike, highlighting the innovative potential of virtual reality and outlining a plan for its incorporation into the travel industry.

SUGGESTIONS

1. Travel agencies should invest in affordable virtual reality systems to ensure greater accessibility for a wide range of audiences, especially for people with disabilities and financial constraints.
2. Artificial intelligence (AI) should be used in future VR tourism applications to personalize virtual experiences according to users' interests, travel experiences, and preferences.

REFERENCES

Beck, J., Rainoldi, M. and Egger, R. 2019. Virtual reality in tourism: A state-of-the-art review. *In*: Tourism Review. 74(3): 586–612. Emerald Group Holdings Ltd. https://doi.org/10.1108/TR-03-2017-0049

Bekele, M.K., Pierdicca, R., Frontoni, E., Malinverni, E.S. and Gain, J. 2018. A survey of augmented, virtual, and mixed reality for cultural heritage. Journal on Computing and Cultural Heritage (JOCCH). 11(2): 1–36. https://doi.org/10.1145/3145534

Buhalis, D. (n.d.). Tourism Management Strategic use of information technologies in the tourism industry Strategic Use of Information Technologies in the Tourism Industry.

Buhalis, D., Harwood, T., Bogicevic, V., Viglia, G., Beldona, S. and Hofacker, C. 2019. Technological disruptions in services: Lessons from tourism and hospitality. Journal of Service Management. 30(4): 484–506. https://doi.org/10.1108/JOSM-12-2018-0398

Chang, K.Y. and Ku, E.C. 2023. Virtual reality technology as a marketing catalyst for travel agencies. Journal of Quality Assurance in Hospitality and Tourism. 1–20.

Cheong, R. 1995. The virtual threat to travel and tourism. Tourism Management. 16(6): 417–422. https://doi.org/10.1016/0261-5177(95)00049-T

Florek, M. and Lewicki, M. 2022. Destinations, virtual reality and COVID-19. How isolation has shaped the behaviors and attitudes towards VR. Economics and Sociology. 15(1): 205–221.

Freina, L. and Ott, M. 2015. A literature review on immersive virtual reality in education: state of the art and perspectives. eLearning and Software for Education. (1): 133–141.

Guttentag, D.A. 2010. Virtual reality: Applications and implications for tourism. Tourism Management. 31(5): 637–651. https://doi.org/10.1016/j.tourman.2009.07.003

Han, D.-I. D., Jung, T. and Gibson, A. 2019. Dublin AR: Implementing augmented reality (AR) in tourism. *In*: Jung, T. and tom Dieck, M.C. (eds). Augmented Reality and Virtual Reality. 89–102. Springer. https://doi.org/10.1007/978-3-030-06246-0_7

Heilig, M.L. 1962. Sensorama Simulator. U.S. Patent No. 3,050,870. https://www.globenewswire.com/news-release/2024/11/13/2980386/0/en/Virtual-Reality-in-Tourism-Market-Report-2024-Compound-Annual-Growth-Rate-of-38-Forecast-During-2024-2028-with-Market-Set-to-Reach-19-91-Billion-in-2028.html

https://www.gminsights.com/industry-analysis/virtual-tourism-market

Ivanov, S.H., Webster, C. and Berezina, K. 2017. Adoption of robots and service automation by tourism and hospitality companies. Revista Turismo and Desenvolvimento. 27(28): 1501–1517.

Jung, T.H. and tom Dieck, M.C. 2017. Augmented reality, virtual reality and 3D printing for the co-creation of value for the visitor experience at cultural heritage places. Journal of Place Management and Development. 10(2): 140–151. https://doi.org/10.1108/JPMD-07-2016-0045

Kostyk, A. and Sheng, J. 2023. VR in customer-centered marketing: Purpose-driven design. Business Horizons. 66(2): 225–236.

Lanier, J. 2001. Virtually there. Scientific American. 284(4): 66–75. doi: 10.1038/scientific american0401-66.

Lavuri, R. and Akram, U. 2024. Role of virtual reality authentic experience on affective responses: moderating role virtual reality attachment. Journal of Ecotourism. 23(3): 347–365.

Leung, W.K., Cheung, M.L., Chang, M.K., Shi, S., Tse, S.Y. and Yusrini, L. 2022. The role of virtual reality interactivity in building tourists' memorable experiences and post-adoption intentions in the COVID-19 era. Journal of Hospitality and Tourism Technology. 13(3): 481–499.

Lundgren, R., and Lööv, J. (n.d.). Virtual Reality-a solution for sustainable travel? A quantitative study of Virtual Reality as a means for environmentally sustainable travel.

Manoharan, G., Rao, C.G., Ashtikar, S.P., Kumar, S. and Nivedha, M. 2024. Voyage virtuoso: Artificial intelligence in transforming tourism. *In*: Utilizing Smart Technology and AI in Hybrid Tourism and Hospitality. 79–97. IGI Global.

Maples-Keller, J.L., Bunnell, B.E., Kim, S.J. and Rothbaum, B.O. 2017. The use of virtual reality technology in the treatment of anxiety and other psychiatric disorders. Harvard Review of Psychiatry. 25(3): 103–113.

Mirzaei, R., Chakerreza, S. and Tabrizi, N. 2024. The role of virtual reality in tourism marketing: Virtual destination image formation and behavioral intentions. International Journal of Hospitality and Tourism Administration. 1–23. https://doi.org/10.1080/1525 6480.2024.2386702

Morrison, A.M., Batabyal, D., Rawal, Y.S. and Sinha, R. 2024. The effects of Bollywood film-induced tour guiding on revisit intentions and destination image. Journal of Vacation Marketing. 13567667241293793.

Oncioiu, I. and Priescu, I. 2022. The use of virtual reality in tourism destinations as a tool to develop tourist behavior perspective. Sustainability. 14(7): 4191.

Rainoldi, M., Driescher, V., Lisnevska, A., Zvereva, D., Stavinska, A., Relota, J., et al. 2018. Virtual reality: An Innovative Tool in Destinations' Marketing. The Gaze: Journal of Tourism and Hospitality. 9: 53–68. https://doi.org/10.3126/gaze.v9i1.22883

Shukla, V., Rana, S. and Prashar, S. 2024. Examining the potential of virtual and augmented reality in enhancing tourism experiences. The Bottom Line.

Sinha, R., Batabyal, D., Bagchi, G. and Selem, K.M. 2023. Establishing the nexus between library and tourism: An empirical approach. Journal of Quality Assurance in Hospitality and Tourism. 1–4. https://doi.org/10.1080/1528008X.2023.2272632

Sousa, N., Jorge, F., Teixeira, M.S., Losada, N., Alén, E. and Guttentag, D. 2024. Does technological innovativeness influence users' experiences with virtual reality tourism? International Journal of Tourism Research. 26(4): e2730.

Tussyadiah, I.P., Wang, D., Jung, T.H. and tom Dieck, M.C. 2018. Virtual reality, presence, and attitude change: Empirical evidence from tourism. Tourism Management. 66: 140–154.

Tussyadiah, I.P., Wang, D., Jia, C. 2017. Virtual reality and attitudes toward tourism destinations. *In*: Schegg, R. and Stangl, B. (eds). Information and Communication Technologies in Tourism 2017. Springer, Cham. https://doi.org/10.1007/978-3-319-51168-9_17

Ukpabi, D., Karjaluoto, H., Olaleye, S.A. and Mogaji, E. 2018. Dual perspectives on the role of artificially intelligent robotic virtual agents in the tourism, travel and hospitality industries. *In*: Vrontis, D., Weber, Y. and Tsoukatos, E. (eds). Proceedings of the 11th Annual Conference of the EuroMed Academy of Business (EMAB) (pp. 1339-1351). EuroMed Press. EuroMed Academy of Business Conference Book of Proceedings. http://www.emrbi.org/bop2018.pdf

Wei, W. 2019. Research progress on virtual reality (VR) and augmented reality (AR) in tourism and hospitality: A critical review of publications from 2000 to 2018. Journal of Hospitality and Tourism Technology. 10(4): 539–570.

Yersüren, S. and Özel, Ç.H. 2024. The effect of virtual reality experience quality on destination visit intention and virtual reality travel intention. Journal of Hospitality and Tourism Technology. 15(1): 70–103.

Zlatanov, S. and Popesku, J. 2019. Current applications of artificial intelligence in tourism and hospitality. *In*: Proceedings of the 11th Annual Conference on Information Technology and Data Related Research (Sinteza 2019). pp. 84–90. https://doi.org/10.15308/Sinteza-2019-84-90

Chapter 9

Smart Kitchens in Luxury Hotels: Implementation and Future Trends

Ricky Gupta
Assistant Professor, JECRC University, Jaipur

SMART KITCHEN OVERVIEW: MACHINE LEARNING PAST, PRESENT, FUTURE

There were times when the kitchens were working in the most orthodox manner, which lacked productivity. According to Buhalis and Leung (Buhalis and Leung 2018), technology implementations in those days were frequently crude, confined to standalone point-of-sale (POS) software or simple countertop displays, and lacked networked intelligence.

Modern technology is being progressively incorporated into the luxury hotel industry to transform culinary procedures and improve the visitor experience. With the integration of artificial intelligence (AI), machine learning (ML), as well as the internet of things (IoT) and data-oriented analytics, smart kitchens have become a game-changer among these advances. They optimize kitchen processes, minimize food waste, and promote susstainability efforts. Potential benefits of AI-based food loss monitoring systems in hotel kitchens were emphasized in research by (Sigala et al. 2025).

*For Correspondence: rickybhm1@gmail.com

DOI: 10.1201/9781003540083-9

AI-powered solutions like Winnow Vision, which track food disposal trends using artificial intelligence and deep learning, have been adopted by luxury hotel brands including Hilton Hotels, Mandarin Oriental, as well as Iberostar. For instance, after implementing solutions powered by AI across many locations, Iberostar claimed to have saved 213 tons of food and 533,000 meals (Dive 2023). In a similar vein, Conrad Dubai reduced food waste by 50%, which resulted in 240 tons of carbon dioxide emissions avoided and $88,000 in yearly savings (Allan 2023). Additionally, handling inventory and menu development is being transformed by smart kitchens. In order to assist chefs in customizing meals while reducing excessive production, AI models are trained on previous sales data, guest choices, and seasonal availability.

In addition to meeting premium service standards, this individualized method advances global objectives for environmentally friendly growth and waste reduction (Limna 2022). The increasing popularity of automated methods of cooking, proactive upkeep for kitchen appliances, and energy optimization is also strongly linked to the deployment of smart kitchens. According to IEEE research, kitchen appliances with integrated AI systems may cut energy use by more than 20%, which will help luxury homes save money over time (Golshany et al. 2025). The coming decades of luxury gastronomic services will be greatly influenced by intelligent kitchens as the restaurant and hotel industry keeps having trouble with issues of sustainability, cost optimization, and changing visitor expectations.

(ABODE 2024) With a current valuation of $186 billion, the artificial intelligence (AI) sector is predicted to rise quickly to $826 billion by 2030. AI's development in the hospitality sector is not unexpected given its potential to automate monotonous operations that free up crucial human hours. Indeed, compared to other hotel technologies, 86% of international hotel chains want to invest in AI (Stephens 2025). AI is already changing the administrative processes of restaurants, including accounting and human resources. According to David Yun, the co-owner of C as in Charlie and Kisa in New York, "tasks that typically require specially trained employees—like bookkeeping duties, handling payroll, and compliance monitoring—can be accomplished quickly and with higher accuracy." Figure 9.1 depicts AI in Smart Kitchens: Past, Present, and Future.

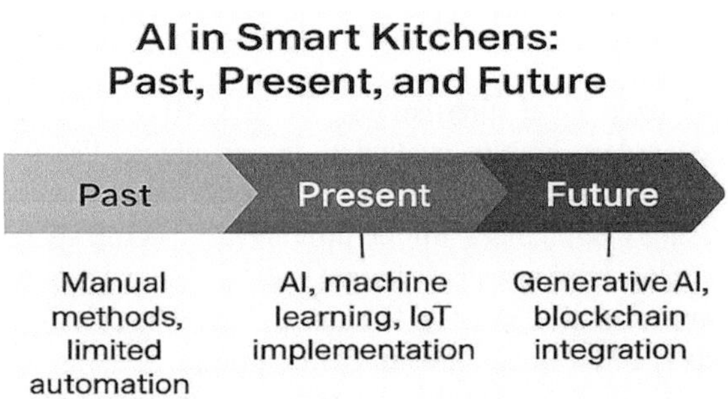

Figure 9.1 AI in Smart Kitchens: Past, Present, and Future (*Source:* Author generated).

TRANSFORMING EXQUISITE HOSPITALITY: SMART KITCHEN INNOVATIONS FOR INCREASED PRODUCTIVITY, ECO-FRIENDLINESS, AND TAILORED VISITOR EXPERIENCES

By integrating the latest innovations like artificially intelligent models (AI), machine learning setups (ML) in synergy with augmented reality (AR) as well as/or virtual reality (VR), smart outlets in upscale properties have the potential to completely transform the hotel and restaurant industry. Enhancing visitor interactions, streamlining processes, and advancing sustainability are the goals of these advances. Figure 9.2 depicts: Modern technology in luxury hotels. The following 12 main advantages show how such technological advancements will influence posh hotel facilities in the subsequent years:

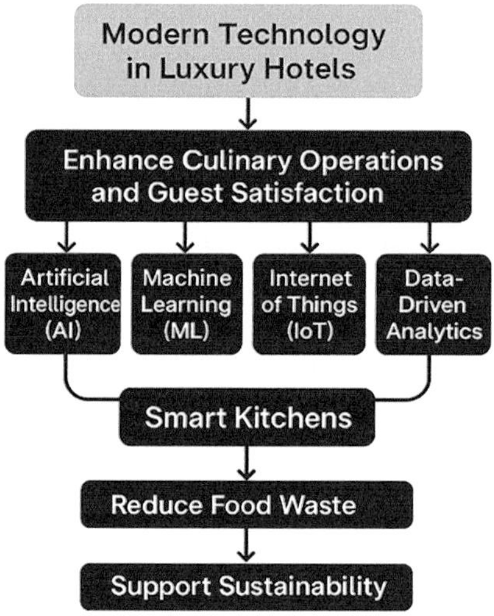

Figure 9.2 Modern technology in luxury hotels (*Source:* Author-generated).

EXCEPTIONALLY TAILORED DINING EXPERIENCES

To provide customized meal alternatives, AI and ML may examine historical behaviors or past orders by the customer or consumer, dietary limitations or preferences, and guest inclinations (Patil 2025). Real-time cooking procedure adjustments by smart appliances might provide a customized meal experience (Thomas et al., 2025). Dining establishment menus that strike a balance between environmental impact and consumer happiness have been designed using Large Language Models (LLMs) in conjunction with optimization approaches. These AI algorithms may recommend menu items that drastically lower emissions related to food choices by evaluating a number of variables. Furthermore, international hotel chains may soon be able to securely exchange information without sacrificing privacy, thanks to shared learning, a decentralized machine learning approach.

This might result in more intelligent, internationally comparable menu refining and tactics to minimize waste (Yang et al. 2019).

PREDICTIVE MANAGEMENT OF ENERGY, REDUCING OVERALL CARBON EMISSIONS

Hotels can foresee trends of energy use by using Artificial Neural Networks (ANN), which enables them to optimize energy use and save money on operational costs (Oscar Trull 2024). By eliminating waste, this proactive capacity to forecast aids initiatives that are aimed at sustainability. This not only helps in reducing the energy consumption but also has something to do with the carbon footprint (Leong and K. 2023). One of the main causes of carbon emissions in the hotel industry is food waste. Continuous monitoring and evaluation of food waste trends has been made possible by the implementation of AI innovations, such as CVS (computer vision systems). According to research, automating these tasks may save operational expenses by about 40% and increase precision by up to 95% (Chui et al. 2016). These technologies enable focused remedies by determining the origins and amounts of waste, which significantly lowers food waste and related emissions. To better manage utilities, smart kitchens are also using AI to forecast peak consumption periods and optimize water consumption in dish-cleaning machines (Gabriele Bartolozzi 2024).

Intelligent Appliances Powered by Internet of Things (IoT): Intelligent ovens and freezers are examples of linked kitchen appliances that can be operated remotely, help in keeping a check on inventory, and recommend meals depending on the commodities that are available in the current stock. Food waste is decreased while productivity is increased by these accessibility features that are embedded in such appliances (Mei, How AI is Shaping the Future of Global Development: Key Insights from CID's Global Empowerment Meeting 2025).

INTERFACES WITH VOICE INPUT FOR OPERATING AND MANAGEMENT

(BlogTeam 2025) Speech assistant connectivity makes it possible to operate kitchen equipment hands-free, enhancing productivity and sanitation for both employees and visitors. Conversational commands are capable of controlling temperatures, illumination, and even recipes.

CULINARY DEVELOPMENT AND TRAINING AUGMENTED BY AR

(Patil 2025) Training for employees may be facilitated and uniformity in food preparations ensured by using augmented reality to show engaging, step-by-step culinary guidelines onto surfaces in the kitchen. AR-based culinary training has been found to boost retention of information by up to 30%, giving personnel engaging, real-time directions that assure uniformity in food creations (Gonzalez-Franco and Lanier 2017)

Using VR to Plan and Design Kitchens: Prior to actual deployment, management of hotels may optimize productivity and space utilization by using virtual environment to see and alter kitchen setups in a simulated setting (Patil 2025). Prior to the implementation of actual settings, virtual reality (VR) tools are being used to pre-visualize the kitchen's functional area layouts, simplifying space efficiency and functional flow (Rauschnabel et al. 2019).

Non-Contact Technology for Improved Hygiene: (Kamal 2024) Voice-activated appliances, sensor-operated garbage cans, and motion-activated plumbing all lessen physical touch and encourage a clean atmosphere, which is important given current global health concerns.

SUSTAINABLE INITIATIVES DRIVEN BY DATA

AI may evaluate ingredient sourcing along with supply channels to find local procurement options, lowering carbon footprints and boosting local economies (Ashley 2025). These methods, which are fueled by data, are in line with the increased awareness of environmental concerns (Yudhistira, et al. 2024) Artificial Intelligence is being used in the culinary sector to optimize energy-intensive food processing techniques like heat drying. The environmental impact of food production is decreased by using AI algorithms to calculate the ideal drying settings, which improve energy efficiency and preserve the quality of the food.

Flexible Kitchen Setups: (Mei, How AI is Shaping the Future of Global Development 2025) In order to provide an ideal working environment and power efficiency, smart systems may modify airflow, temperature, and illumination in response to real-time preparation activity.

Tailored Visitor Experiences: Systems powered by AI have the ability to instantly recall visitor preferences, enabling personalized welcome messages, food recommendations, and dining experiences that increase patron loyalty and happiness (BlogTeam 2025). Chefs may now make ecologically responsible decisions by using AI-aided life cycle assessments (LCAs) to get real-time data on the environmental effects of certain meals (N.R. Nawani 2021).

Simplified Inventory Control: Real-time inventory level tracking by IoT devices lowers the possibility of shortages or overstocking by automatically reordering goods as needed. Smart chillers equipped with AI-powered computer vision technologies are able to manage stocks, identify degradation, and continually monitor inventories. By enabling prompt actions, this real-time analysis lowers food wastage and guarantees freshness (Castillo-Ortiz and Villar-Patiño 2024). This will meet the growing demand from guests for environmental sustainability and accountability by guaranteeing not just traceability but also validation of organic certification as well as ethical procurement (Galvez et al. 2018). Comprehensive inventory item tracking across the supply chain is made possible by combining artificial intelligence (AI) with Internet of Things (IoT) devices (Dhal and Kar 2025). Transparency is guaranteed, traceability is enhanced, and prompt answers to inventory inconsistencies are made possible by this connection. Demand estimation technologies powered by AI, for example, have improved procurement procedures, decreasing overordering and increasing profits (Li et al. 2021). Product spoilage is being decreased and supply chain traceability is being improved by IoT-enabled chilling units and sensor-supported store monitoring (Jagtap et al. 2022).

Personalized Dining Situations: Fancy dining may take on a new level with the help of AR and VR technology, which can be used to create interactive dining spaces that let visitors enjoy virtual culinary adventures or customized settings (HFTP 2024). A study published in the International Journal for Multidimensional Research Perspectives emphasizes how AI is revolutionizing hotel dining. AI-driven algorithms evaluate visitor information to provide personalized eating experiences and suggestions (Milton 2024). AI algorithms also enhance culinary inventiveness and visitor delight by optimizing menu options and suggesting distinctive meals (Yang et al. 2024). A new personalized recommender system that uses Aspect-Based Sentiment Analysis (ABSA) is presented in research that was published in Tourism Management. The technology determines particular preferences across a range of restaurant elements by examining client evaluations, allowing for more specialized and effective eating recommendations. This method captures complex consumer attitudes better than standard models. Adding this technology to the facilities of upscale hotels improves sustainability as well as operational effectiveness while also improving the visitor experience. Adopting smart kitchen technologies will be essential to satisfying the demands of contemporary tourists as the hospitality sector develops.

CONCLUSION: PAST, PRESENT, AND FUTURE TRENDS IN THE SMART KITCHENS

A significant change in the way gastronomic procedures are planned and carried out is represented by the incorporation of AI, ML, AR, and VR in luxurious lodging facilities. In addition to transforming food processes, the shift from typical orthodox settings to effectively smart kitchens has accelerated industry initiatives towards cost optimization, environmental sustainability, and detailed personalization. Hotel kitchens in the ***past*** scenarios used to run mostly on chef instincts and manual processes with little use of technology assistance. Significant operational problems included poor resource utilization, variable level of service, and food spoilage.

The majority of employee training was based on vocational simulations, and the guest/visitor experiences were uniform instead of customized. The hotel sector is changing dramatically in the ***present*** scenario. Large amounts of administrative and visitor-related data are getting analyzed using AI and ML to provide specialized, effective, and sustainable solutions. Such technological advances are reflected in automatic food waste surveillance, efficient menu planning, continuous inventory administration, and energy usage forecasting.

A particularly noteworthy trend is the expanding usage of AR and VR to boost training for employees as well as engagement with visitors. Engaging virtual dining settings are providing improved, unforgettable experiences that surpass the conventional hospitality structure. AI's capacity to detect carbon-intensive components, optimize high-energy cooking methods, and recommend low-impact substitutes is proving crucial from a sustainability standpoint.

Furthermore, AI's back-of-house managerial solutions are boosting productivity in previously unheard-of ways. AI-powered technologies have simplified tasks like generating pay slips and cheques, compliance monitoring, and personnel scheduling, lowering human errors and saving a significant amount of operational time. Employees may then refocus on an innovative approach and customer service, two important aspects of premium hospitality.

The possibilities of AI and intelligent kitchen technology are only going to grow in the future. Voice-based culinary aides may eventually be able to respond in real-time to chef preferences and guest comments thanks to developments in generative artificial intelligence (AI) and natural language processing (NLP). These systems may support many languages, modify recipes quickly, or even suggest allergen-free substitutes at any time.

Additionally, blockchain will be essential to ethical food procurement when combined with AI and IoT. In summary, AI, ML, IoT, AR, and VR-powered intelligent kitchens are more than simply technological innovations. They represent a comprehensive reimagining of premium hospitality processes, integrating ecological consciousness, guest-centric customization, and data-driven decisions in a streamlined manner. Premium hotels can remain at the vanguard of sustainability and culinary excellence by taking lessons from the past, excelling in the present, and actively planning for upcoming advances.

In a global marketplace that is growing more digital and ecologically concerned, using these technologies is becoming more than simply a competitive advantage.

REFERENCES

ABODE. 2024. AI in the hospitality industry: 15 real-world examples. Retrieved from Abode Worldwide: Strategic Public Relations. https://abodeworldwide.com/ai-hospitality-industry/

Allan, H. 2023. Conrad Dubai cuts food waste by 50% using AI. Winnow Solutions. Retrieved April 10th, 2025 from https://blog.winnowsolutions.com/conrad-dubai-cuts-food-waste-50

Ashley, D. 2025. Inside the sustainability retreat that's reshaping hotel kitchens worldwide. Retrieved from Food and Wine. https://www.foodandwine.com/marriott-luxury-collection-mad-academy-sustainability-11695866?utm_source

BlogTeam, D.S. 2025. Digital Solley. Retrieved from Top 15 Technology Trends in Hospitality in 2025. https://digitalsolley.com/2025/01/26/top-15-technology-trends-in-hospitality-in-2025/

Buhalis, D. and Leung, R. 2018. Smart hospitality—Interconnectivity and interoperability towards an ecosystem. International Journal of Hospitality Management. 71: 41–50. doi: https://doi.org/10.1016/j.ijhm.2017.11.011

Castillo-Ortiz, I. and Villar-Patiño, C. 2024. Computer vision solution for uniform adherence in gastronomy schools: An artificial intelligence case study. International Journal of Gastronomy and Food Science. 37. doi: https://doi.org/10.1016/j.ijgfs.2024.100997

Cheng, I.K. and Leong, K.K. 2023. Data-driven decarbonisation pathways for reducing life cycle GHG emissions from food waste in the hospitality and food service sectors. Scientific Reports. 13(1): 418. doi: https://doi.org/10.1038/s41598-022-27053-6

Chui, M., Manyika, J. and Miremadi, M. 2016. Where machines could replace humans—and where they can't (yet). McKinsey Quarterly. Retrieved from https://www.mckinsey.com/capabilities/mckinsey-digital/our-insights/where-machines-could-replace-humans-and-where-they-cant-yet

Dhal, S.B. and Kar, D. 2025. Leveraging artificial intelligence and advanced food processing techniques for enhanced food safety, quality, and security: a comprehensive review. Discover Allied Sciences. 75(7). Retrieved from http://link.springer.com/article/10.1007/s42452-025-06472-w

Dive, H. 2023. Iberostar saves 533,000 meals with AI kitchen tech. Retrieved from https://www.hoteldive.com/news/ai-assisted-kitchen-solution-to-reduce-food-waste/649553/

Gabriele Bartolozzi, G.P. 2024. Energy saving starts in the kitchen. Energy and Buildings. 322: 114726. doi: https://doi.org/10.1016/j.enbuild.2024.114726

Galvez, J.F., Mejuto, J.C. and Simal-Gandara, J. 2018. Future challenges on the use of blockchain for food traceability analysis. Trends in Analytical Chemistry. 107: 222–232. doi: https://doi.org/10.1016/j.trac.2018.08.011

Golshany, H., Ni, Y., Yu, Q. and Fan, L. 2025. IoT-enabled smart kitchen technologies and their impact on food storage, preparation, and culinary experiences: A systematic review. Food Research International. doi: https://doi.org/10.1016/j.foodres.2025.116557

Gonzalez-Franco, M. and Lanier, J. 2017. Model of illusions and virtual reality. Frontiers in Psychology. 8: 1125. doi: https://doi.org/10.3389/fpsyg.2017.01125

HFTP, T. 2024. Technology Trends Shaping the Future of the Hospitality Industry Today. Retrieved from HFTP. https://www.hftp.org/news/4124813/technology-trends-shaping-the-future-of-the-hospitality-industry-today?utm_source

Jagtap, S., Rahimifard, S. and Gupta, S. 2022. IoT and AI for sustainable food supply chains: A review. Journal of Cleaner Production. 343: 130987. Retrieved from https://www.sciencedirect.com/science/article/abs/pii/S1466856425001031

Kamal, Q. 2024. Kitchen Trends Shaping the Hospitality Sector in 2024. Retrieved from Elite Edge Resources. https://eliteedgeresources.com/kitchen-trends-shaping-the-hospitality-sector-in-2024/

Li., L., Fang, Y., Lim, A. and Wang, X. 2021. Demand forecasting for the food service industry using machine learning. Computers and Industrial Engineering. 157: 107334. Retrieved from https://www.sciencedirect.com/science/article/pii/S2212827119301568 https://doi.org/10.1016/j.procir.2019.02.042

Limna, P. 2022. Artificial Intelligence (AI) in the Hospitality Industry: A Review Article. International Journal of Computing Sciences Research. 1306–1317. doi:https://doi.org/10.25147/ijcsr.2017.001.1.103

Mavani, N.R., Ali, J.M., Othman, S., Hussain, M.A., Hashim, H. and Rahman, N.A. 2021. Application of Artificial intelligence in food industry—A guideline. Food Engineering Review. 14(1): 134–175. doi:https://doi.org/10.1007/s12393-021-09290-z

Mei, M. 2025. How AI is Shaping the Future of Global Development: Key Insights from CID's Global Empowerment Meeting. Retrieved from Harward Center for International Development: https://www.hks.harvard.edu/centers/cid/voices/how-ai-shaping-future-global-development-key-insights-cids-global-empowerment

Milton, T. 2024. Artificial Intelligence Transforming Hotel Gastronomy: An In-depth Review of AI-driven Innovations in Menu Design, Food Preparation, and Customer Interaction. (D.A. Singh, Ed.). International Journal for Multidimensional Research Perspectives. 47–61. Retrieved from https://www.chandigarhphilosophers.com/index.php/ijmrp/index

Patil, R. 2025. What's Next for Smart Kitchens? Exploring the Future of Kitchen Technology. Retrieved April 07, 2025, from Businessoffood.in: https://www.businessoffood.in/whats-next-for-smart-kitchens-exploring-the-future-of-kitchen-technology/

Rauschnabel, P., Felix, R. and Hinsch, C. 2019. Augmented reality marketing: How mobile AR-apps can improve brands through inspiration. Journal of Retailing and Consumer Services. 49: 43–53. doi: https://doi.org/10.1016/j.jretconser.2019.03.004

Sigala, E.G., Gerwin, P., Chroni, C., Abeliotis, K., Strotmann, C. and Lasaridi, K. 2025. Reducing food waste in the HORECA sector using AI-based waste-tracking devices. doi: https://doi.org/10.1016/j.wasman.2025.02.044

Stephens, R. 2025. How AI is Revolutionizing Restaurant Kitchens. Retrieved from Food and Wine Magazine: https://www.foodandwine.com/restaurants-ai-artificial-intelligence-technology-11683792

Thomas, A.T., Yee, A., Mayne, A., Maya B. and Mathur, D.J. 2025. What can large language models do for sustainable food? doi: https://doi.org/10.48550/arXiv.2503.04734

Trull, O., Peiro-Signes, A., Garcia-Diaz, J.C. and Segarra-Ona, M. 2024. Prediction of energy consumption in hotels using ANN. Retrieved April 08, 2025, from https://arxiv.org/abs/2405.18076: https://doi.org/10.48550/arXiv.2405.18076

Yang, Q., Liu, Y., Chen, T. and Tong, Y. 2019. Federated machine learning: Concept and applications. ACM Transactions on Intelligent Systems and Technology (TIST). 10(2): 1–19. doi: https://doi.org/10.1145/3298981

Yang, S., Jang, D., Kim, J. and Li, Q. 2024. Deep learning mechanism and big data in hospitality and tourism: Developing personalized restaurant recommendation model to customer decision-making. Tourism Management. 121. doi: https://doi.org/10.1016/j.ijhm.2024.103803

Yudhistira, B., Adi, P., Mulyani, R., Chang, C.-K., Gavahian, M. and Hsieh, C.-W. 2024. Achieving sustainability in heat drying processing: Leveraging artificial intelligence to maintain food quality and minimize carbon footprint. Comprehensive Reviews in Food Science and Food Safety. 23. doi: https://doi.org/10.1111/1541-4337.13413

Chapter **10**

Revolutionizing Hospitality: The Role of Blockchain in Driving Transparency, Security, and Efficiency

Shikha*[1] and Yashwant Singh Rawal[2]
[1]Assistant Professor, Central University of Haryana, Mahendergarh
[2]Professor, Amity University Rajasthan, India

INTRODUCTION

Despite its recent introduction, blockchain innovation in the movement sector is still in the early stages of development. The transportation and travel industry is gaining a lot of attention from scientists from every corner of the globe, pursuing systems and applications involving blockchain. The use of blockchain has grown in acceptance over the last few years, resulting in new and useful uses in the hospitality industry (Dhiraj et al. 2023). As a result of today's dynamic marketplace, the hospitality sector faces challenges that pose threats to profitability and operational efficiency. Booking cancellations and fraudulent transactions account for 10% of all bookings, resulting in significant revenue erosion. Additionally, intermediaries collect a commission of up to 20%, putting additional strain on the household's finances. Blockchain presents itself as a promising technology due to its decentralized nature. Furthermore, it promises

*For Correspondence: shikhahospitality@gmail.com, yashwantr84@gmail.com

DOI: 10.1201/9781003540083-10

reduced middleman dominance in the ecosystem, in addition to data integrity (Singh 2023). With blockchain technology, hotel front office transactions can be more secure, providing a robust solution for many challenges they face. Guests' sensitive information can be better protected because it is decentralized and immutable, operational processes can be streamlined, and transparency is increased, which ultimately leads to more trust and satisfaction among guests (Manikkuwar and Bagade 2024). A number of aspects of a hotel's business can be enhanced by implementing blockchain-based solutions, including increased efficiency, cost reduction, greater transparency, and increased security (Rawat and Poonam 2023).

PURPOSE OF THE STUDY

An assessment of how blockchain technology can improve transparency, increase security, and organizational efficiency in the hospitality industry is the purpose of this study. The goal is to understand how blockchain can be used to solve problems, as well as to understand its challenges and future in the hospitality industry.

LITERATURE REVIEW

What is Blockchain?

Today's economy relies heavily on new and upcoming technologies, such as blockchains. All transactions or digital events are recorded in a blockchain, which is a distributed ledger shared among participants. It facilitates the transfer of value within a business network by providing provenance, immutability, and finality. In addition to enabling real-time value exchange, it reduces errors and costs. Through cryptography, trust can be established between parties involved in a transaction using a network consensus approach (Saha et al. 2019).

The hospitality industry is dynamic, and blockchain technology can help hotels eliminate intermediaries and third parties. The digital age has made data the new currency, so staying safe as a hotel guest is crucial in this digital age. Having it at your hotel can boost its performance and help it stand out from the competition (Lacalle 2023).

Blockchain Types and Characteristics (Table 10.1)

In Blockchain Technology, there are different categories based on the access control mechanisms, governance models, and how the data is verified or stored. A blockchain can be used for a wide variety of applications depending on its unique features. Blockchains may be public, private, or consortium, with each having its own benefits. In addition, there is a fourth type of potential blockchain termed Hybrid Blockchain (Fischer and Arnold 2024).

Table 10.1 Blockchain types and their characteristics

Blockchain Type	Decentralization	Access Control	Transparency	Use Cases	Example Blockchain
Public blockchain	Fully decentralized	Open (Anyone can participate)	Full transparency	Cryptocurrencies, DeFi	Bitcoin, Ethereum
Private blockchain	Centralized or semi-centralized	Permissioned (selected members)	Restricted transparency	Supply chain, Banking, Healthcare	Hyperledger Fabric, Ripple
Consortium blockchain	Semi decentralized	Mixed (Public/Private)	Partial	Financial Institutions, Cross-border payments	R3 Corda, IBM Food Trust
Hybrid blockchain	Customizable		Customizable	Government Enterprises	Dragonchain, XinFin

Source: Abbas 2024

How Blockchain Works?

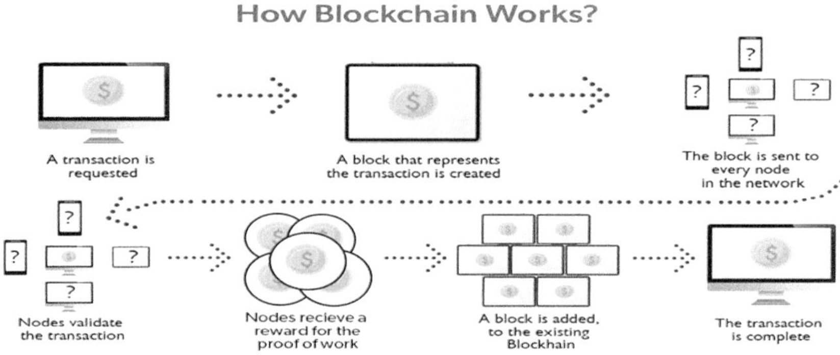

Figure 10.1 How blockchain works? (*Source:* https://www.geeksforgeeks.org/how-does-the-blockchain-work/).

As per the Figure 10.1 the blockchain network receives a new transaction. A double encryption process is performed using a public and private key. As soon as the transaction has been sent, it is sent to the global network of peer-to-peer computers. Each node checks if there is enough money to be able to complete the transaction by checking the validity of the transaction (Arpit 2025). Typically, blockchain networks consist of many nodes, and many transactions are verified simultaneously. In the case that a transaction has been confirmed and declared legitimate, it will be added to the mempool as soon as it has been verified. In a block, multiple mempools are combined to form all the verified transactions at one node. A block will be made permanent by adding it to the blockchain network. However, if all nodes can add blocks in this manner, it will disrupt the blockchain network's operation. This problem is resolved by a consensus mechanism, which ensures that every new block added to the (GeeksforGeeks 2024). Blockchain represents the only version of truth that is agreed upon by all the nodes, and that only valid blocks are securely attached to the Blockchain. In the blockchain, nodes are rewarded for adding blocks to the blockchain. Blockchains are built by adding blocks to them with a hash code generated by the consensus algorithm. Adding the newly created block to the blockchain will occur once the hash value and authenticity of the block have been determined (Grasso 2023). A blockchain is formed by cryptographically linking blocks using their hashes from previous blocks. Each time the blockchain is updated, a new block is added. In addition to completing the transaction, the blockchain also stores the details of the transaction forever. The details of the transaction can be retrieved and confirmed by anyone (Shalini 2025).

Blockchain Technology and the Hospitality Industry

A blockchain technology allows multiple parties to manage data without the need for an intermediary, and it does so without requiring the creation of trust between the parties. Distributed ledger technology can be used to control data in a decentralized manner, ensuring its immutability and allowing retrospective checks. A variety of industries, including tourism, are undergoing transformations due to it. It has become increasingly popular among businesses and individuals. As blockchains are able to transmit value through their data stores, they can be used to complete financial transactions, such as payments and transfers, without the use of intermediaries (BEŞTAŞ 2024). There are endless benefits to be gained from the use of blockchain technology from an economic, administrative, and service standpoint. Both private and public sectors will benefit from it in terms of saving time and money associated with economic and financial operations. A blockchain also allows the monitoring of all operations and the verification of the source, as well as addressing fraudulent activities, forgeries, manipulations, and tampering. Using blockchains will improve the quality of customer service, loyalty, and profitability of tourism and hospitality entities due to encryption, fragmentation, and complete user security (Darwesh 2023). Hospitality businesses are often affected by high operating costs and inefficiencies caused by the use of intermediaries such as Online Travel Agencies (OTAs) and Global Distribution Systems (GDS) (Batabyal and Das 2013). As a result of blockchain technology, service providers and customers can transact directly, eliminating these intermediaries. Besides reducing costs, this improves the customer experience by facilitating faster transactions and improving transparency. The use of blockchain-based booking platforms, for instance, allows customers to engage directly with hotels and airlines without being charged inflated fees (Anil and Bhartwal 2024). Blockchain technology has the potential to improve food waste management in the hospitality sector. Through increased transparency in food supply chains, it is possible to track food problems from production to consumption, as well as prevent food waste, fraud, and other unethical practices and actions. Furthermore, it enables customers and consumers to make better eco-conscious choices by educating them about the way food is developed and delivered (Stroumpoulis et al. 2022). With blockchain technology, you can perform a number of real-world functions as well as save time, effort, and money, and it can be used to monitor all operations and verify their sources, as well as prevent fraud, forgery, manipulation, and tampering due to its encryption and hashing mechanisms. Blockchain technology can help hotels, travel agencies, airlines, restaurants, and other hospitality companies provide better service, attract more customers, and enhance profitability (Abdelkawi and Salama 2022).

Benefits of Blockchain for the Hospitality Industry

1. ***Safety:*** Due to the abundance of easily accessible data, security risks are increasingly plaguing the hospitality sector. All data is traceable and decentralized thanks to blockchain technology. Data cannot be erased or taken down, not even in the event of a cyberattack. The reputation of your hotel depends on your ability to protect guest information (Lacalle 2023).
2. ***Easier Service Payment:*** Thanks to blockchain technology, both you and your guests can be confident that the payment method they have selected

will be accepted. This is because the entire process is more transparent and secure, as all payments are handled globally. Learn why providing various payment choices is a good idea for hotels (Barten 2024).

3. ***Decreased Dependence on Outside Parties:*** Reservations may now be tracked and controlled centrally thanks to technological advancements. Direct reservation booking by visitors eliminates the need for middlemen. This lowers transaction fees associated with payment processing and commissions. Better management of hotel inventory, more flexible booking systems, and improved payment and reservation processing are further advantages (Pandiyan et al. 2023).
4. ***Improved Efficiency:*** Blockchain technology can enhance efficiency in various hotel industry processes, including cancellations, payments, and bookings. As a result, intermediaries will be eliminated, costs will be reduced, and efficiency will be improved (Rawat and Poonam 2023).
5. ***Prevent Duplicate Bookings:*** Duplicate bookings can be prevented with blockchain implementation. During the holiday and vacation seasons, when centralized payment servers are heavily loaded, there is a tendency for users to book multiple rooms. As a result of centralization, multiple bookings of the same entity may occur. As a decentralized technology, blockchains are resistant to single points of failure or denial-of-service attacks (Khanna et al. 2020).

Uses of Blockchain in the Hospitality Industry

1. ***Confirming One's Identity:*** Given the numerous opportunities for identity fraud, thoroughly confirming the identities of your guests is crucial to maintaining the security of both your hotel and its patrons. Blockchain enables businesses to use a dispersed network of computers to confirm identity. Because guests can safely transmit check-in paperwork in advance of their arrival, check-in times are shortened (Bansal 2023).
2. ***Data Accessibility:*** In addition to providing hotels with a secure way to store data, blockchain will also improve customer satisfaction and efficiency by providing a single spot for customers to view and access data. In addition to tracking the check-in and check-out times of guests, hotels are able to control all aspects of operational management (Sharma 2021).
3. ***Customization:*** In addition to customizing the guest experience, blockchain technology can be used to enhance the hotel experience by knowing the preferences of hotel guests. Before guests check in to their rooms, hotels could set the rooms' temperature and lighting to their preferences. In-room amenities can be tracked while they are being used, and connected sensors can be integrated with blockchain systems to ensure everything is working (Sanghi 2023).
4. ***Room Access and Security:*** Hotel keys can be managed through smart contracts implemented using blockchain technology. Blockchain

technology can be used for storing authorized digital IDs and payment accounts, which makes reserving rooms and getting keys quick and easy. Hotels can manage their keys more efficiently and securely with this system as it ensures transactions are secure without requiring physical keys (Barkel et al. 2021).

5. ***Loyalty Programs:*** Loyalty programs that are based on blockchain technology can be more efficient and secure. The use of blockchain technology can help hotels create a loyalty program that is secure and transparent, as well as unable to be manipulated or fraudulent (Rawat and Poonam 2023).

Challenges of Blockchain Technology Adoption in Hospitality Industry

1. ***Lack of Awareness and Understanding:*** Hotel and tourism industry officials are not aware of or do not understand blockchain technology, which is one of the biggest challenges to its adoption. Due to their lack of familiarity with blockchain technology and its potential benefits, many industry players are hesitant to invest in it. Also, due to a lack of knowledge, blockchain solutions cannot be developed or implemented (Bridgewater 2022).
2. ***High Cost:*** Putting blockchain into place in a company is very expensive. Many companies are hesitant to invest in this technology due to its capital-intensive nature. You need to hire blockchain software developers and core blockchain developers if you are implementing blockchain. Substantial funds are needed to accomplish this. Next, you'll need to develop blockchain applications. As well as the software, there are hardware requirements (Biswas 2025).
3. ***Energy Consumption:*** Computing power is essential to the secure operation of blockchain technology and the validation of transactions. Due to this computing power, a high amount of energy is required, which negatively impacts the environment. In addition, blockchain technology consumes a considerable amount of energy, especially from non-renewable sources (Dhawan 2024).
4. ***Interoperability:*** In the blockchain world, interoperability describes how different blockchain networks can communicate and work seamlessly together. The challenge is achieving interoperability between blockchains because each blockchain uses different protocols and standards.
5. ***A Reluctance to Adopt New Technologies:*** Blockchain technology may be hesitantly adopted by established players in traditional industries due to their resistance to change. Some people are cautious about disrupting existing business models, fear losing control, or simply don't fully understand the technology's potential. It can be helpful for organizations to participate in standardization consortiums and working groups aimed at creating blockchain standards (Saxena 2023).

6. ***Scalability:*** There may be a problem with current blockchain networks handling the volume of transactions that a large hotel operation would need. Transaction processing times could be slowed, and real-time service delivery could be compromised by scalability issues (Löcher 2023).

FUTURE SCOPE OF BLOCKCHAIN TECHNOLOGY IN THE HOSPITALITY INDUSTRY

A positive outlook is expected for blockchain technology in the hospitality industry in the near future. It is expected that the adoption of blockchain will accelerate over the next five years as technology advances and awareness increases. By embracing this technology proactively, hotels will be in a position to benefit not only from safeguarding transactions but also from positioning themselves as innovators in an industry that is rapidly changing (Manikkuwar and Bagade 2024).

CONCLUSION

With its potential to eliminate issues like fraud, inefficiency, or over-reliance on intermediaries, blockchain technology is set to transform the hospitality sector. The research cites several areas where efficiency through transparency and data protection can be achieved using blockchain, including booking and payment systems, identity verification, and loyalty programs. The need for trust and security as data compromises increase in today's economy is addressed by the decentralization and immutability features of blockchain. Though blockchain technology has considerable benefits, implementing it in the hospitality domain is still in its infancy due to high costs, environmental concerns, organizational resistance, and scalability problems. With the advancement of emerging technologies and increased awareness, it is anticipated that the diminishing challenges will make wider adoption possible. Changes in the hospitality industry's service delivery, customer interaction, and operational frameworks will likely occur due to advancements in blockchain technology. Hotels able to embrace change and proactively adopt blockchain in their systems during its emergence stand to greatly benefit by safeguarding their sensitive data while reinforcing their brand image as forward-thinking, digitally-led market leaders.

REFERENCES

Abbas, K.A. 2024. Blockchain technology and its issues: Types, applications challenges and future directions. Journal of Information Systems Engineering and Management. 10(17s): 103–116.

Abdelkawi, A. and Salama, M.A. 2022. Applicability of blockchain technology in Egyptian hotels. Journal of Tourism, Hotels and Heritage. 5(1): 63–80. doi: 10.21608/sis.2022.150908.1068

Anil, V.U and Bhartwal, U. 2024. Role of blockchain in hospitality industry. International Journal of Multidisciplinary Research in Science, Engineering and Technology. 7(12): 18133–18138. doi: 10.15680/IJMRSET.2024.0712093

Arpit. 2025. How does blockchain work. https://intellipaat.com/blog/tutorial/blockchain-tutorial/how-does-blockchain-work/

Bansal, R. 2023. Blockchain Technology in Hospitality Industry (Part-56). https://www.linkedin.com/pulse/blockchain-technology-hotel-industry-part-56-roohi-bansal

Barkel, C., Kurgun, H. and Groen, B. 2021. Blockchain in Hospitality amd Tourism Industry. M3 Publishing. https://doi.org/10.5038/9781732127593

Barten, M. 2024. Blockchain Technology and its Uses in the Hospitality Industry. https://www.revfine.com/blockchain-technology-hospitality-industry/

Batabyal, D. and Das, D.K. 2013. Management of travel booking in the light of destination marketing: Evidence from an Alpine Indian State-Sikkim. Indian Journal of Marketing. 43(11): 36–45.

BEŞTAŞ, M. 2024. Evaluation of blockchain technology in the context of tourism sector within the framework of diffusion of innovation theory: Challenges and opportunities. Journal of Information Systems Engineering and Management. 9(4): 30348. https://doi.org/10.55267/iadt.07.15197

Biswas, S. 2025. Disadvantages of blockchain technology. https://cleartax.in/s/disadvantages-of-blockchain.

Bridgewater, J. 2022. Challenges Facing the Travel and Tourism Industry in the Adoption of Blockchain Technology. Medium. https://medium.com/blockchain-innovation-insights/challenges-facing-the-travel-and-tourism-industry-in-the-adoption-of-blockchain-technology17a51b2cb7b2.

Darwesh, S. 2023. Assessing the impact of blockchain technology on guests behavior in five-star hotels using UFAUT model. Journal of Tourism, Hotels and Heritage. 6(2): 27–41. doi: 10.21608/sis.2023.206030.1136

Dhawan, A. 2024. The dark side of blockchain technology: Environmental concerns and mitigation strategies. Journal of Indian Law and Society. https://jilsblognujs.wordpress.com/2024/03/23/the-dark-side-of-blockchain-technology-environmental-concerns-and-mitigation-strategies/

Dhiraj, A., Kumar, S., Rani, D., Grima, S. and Sood, K. 2023. Blockchain Payment Services in the Hospitality Sector: The Mediating Role of Data Security on Utilisation Efficiency of the Customer. Data. 8(8): 123. https://doi.org/10.3390/data8080123

Fischer, T. and Arnold, M. 2024. Blockchainin government services: Enhancing transparency in public administration. Public Administration Review. 14(1): 34–47.

GeeksforGeeks 2024. How does the Blockchain Work? Retrieved on 19 April, 2025 from https://www.geeksforgeeks.org/how-does-the-blockchain-work/

Grasso, L. 2023. How does blockchain work? Here is a simple explanation. https://www.linkedin.com/pulse/how-does-blockchain-work-here-simple-explanation-linda-grasso-pcgee

Khanna, A., Sah, A., Choudhury, T. and Maheshwari, P. 2020. Blockchain Technology for Hospitality Industry. *In*: Themistocleous, M., Papadaki, M. and Kamal, M.M. (Eds). Information Systems. EMCIS 2020. Lecture Notes in Business Information Processing. Vol. 402. Springer, Cham. https://doi.org/10.1007/978-3-030-63396-7_7

Lacalle, E. 2023. Why is blockchain technology useful in the hospitality industry? https://www.mews.com/en/blog/blockchain-for-hotels .

Löcher, S. 2023. The potential application of blockchain technology in the hospitality industry. https://medium.com/@sophie.loecher/the-potential-application-of-blockchain-technology-in-the-hotel-industry 307d3e4b0e74#:~:text=Scalability%20issues%20could%20lead%20to,energy%2Defficient%20alternatives%20is%20important

Manikkuwar, S. and Bagade, S. 2024. The Future of Blockchain Technology in Securing Hotel Front Office Transactions. Journal of Emerging Technologies and Innovative Research. 11: 9120. (c 165–c 185).

Pandiyan, P., Saravanan, S., Usha, K., Kannadasan, R., Alsharif, M.H. and Kim, M.-K. 2023. Technological advancements toward smart energy management in smart cities. Energy Reports. 10: 648–677. https://doi.org/10.1016/j.egyr.2023.07.021

Rawat, S. and Poonam. 2023. Blockchain in hotel business. International Journal of Research in Applied Science and engineering Technology. 11(6): 1785–1788. https://doi.org/10.22214/ijraset.2023.53891

Saha, S., Jana, B. and Poray, J. 2019. A Study on Blockchain Technology. Available at SSRN: https://ssrn.com/abstract=3477373 or http://dx.doi.org/10.2139/ssrn.3477373

Sanghi, A. 2023. Five benefits of blockchain technology in the tourism sector. https://www.financialexpress.com/business/blockchain-five-benefits-of-blockchain-technology-in-the-tourism-sector-3130755/

Saxena, A. 2023. What are the challenges of implementing blockchain in traditional industries. https://www.financialexpress.com/business/blockchain/what-are-the-challenges-of-implementing-blockchain-in-traditional-industries/3044854/

Shalini 2025. How does blockchain work? Advantages and Uses Explained. https://amityonline.com/blog/how-does-blockchain-works.

Sharma, M. 2021. Use of blockchain in the hospitality industry. https://hospitality.economictimes.indiatimes.com/news/speaking-heads/use-of-blockchain-in-the-hospitality-industry/88117728

Singh, D. 2023. Blockchain in Hospitality: Decentralizing the Guest Experience. https://www.debutinfotech.com/blog/blockchain-in-the-hospitality-decentralizing-the-guest-experience

Stroumpoulis, A., Kopanaki, E. and Oikonomou, M. 2022. The impact of blockchain technology on food waste management in the hospitality industry. ENTRENOVA - ENTerprise REsearch InNOVAtion. 7(1): 419–428. https://doi.org/10.54820/CQRJ6465

Chapter 11

Consumer Perception of AI-Driven Customer Service at JW Marriott Mumbai: A Data-Driven Case Study

Ishan Bakshi[1], Hitesh Marwaha[2], Aashish Samuel[3] and Ashish Ahlawat*[4]

[1]Assistant Professor, Maharishi Markandeshwar (Deemed to be University), Mullana, Ambala, Haryana, India-133207

[2]Assistant Professor, School of Computational Sciences, GNA University, Phagwara, Punjab, India-144401

[3]Assistant Professor, School of Hospitality, GNA University, Phagwara, Punjab, India-144401

[4]Assistant Professor, Sanskaram University, Patauda, Haryana, India-124108

INTRODUCTION

The hospitality sector is experiencing a prototype revolution as Artificial Intelligence (AI) is being introduced into customer service processes. Chatbots, virtual concierges, robot concierges, and predictive analytics have revolutionized consumer-to-business interactions with the emergence of AI technologies. The AI-based solutions allow hospitality businesses to provide customized services, enhance operational effectiveness, and provide 24/7 support, leading to customer

*For Correspondence: ashishahlawat1209@gmail.com

DOI: 10.1201/9781003540083-11

satisfaction and loyalty (Ivanov and Webster 2019a). With the introduction of AI, new customer servicing models, grounded on human interfaces so far, are being re-engineered by intelligent systems and automation.

The perceived attitude of the consumer towards AI-based services stands at the forefront of this paradigm shift in the industry. Overall success and effectiveness of the use of artificial intelligence in the hospitality sector heavily rest on the perceptions held and customer reception towards technology-facilitated interactions with such technology. Though convenience, personalization, and response speed are qualities most easily attributed to AI, it is also true that some guests may indeed find such innovations to lack human touch or even be intrusive to some degree. This is due to perceived privacy, security, and the loss of essential human touch in serving, which are areas of concern leading to such a consumer reaction (Tussyadiah et al. 2020). It is imperative that organizations looking forward to leading and finding equilibrium between technology advancement and true human contact must have a clear understanding of the delicate and sophisticated consumer sentiment regarding AI-enabled services.

Meanwhile, online marketing plays a key role in AI-powered services in the hospitality sector. With the help of marketing strategies like social media marketing, search engine optimization, content marketing, and targeted ads, stakeholders in the hospitality sector can effectively convey AI's value propositions to current and potential guests (Das et al. 2024). Additionally, AI is an integral component of digital marketing since machine learning programs enable businesses to tap into vast customer datasets, segment the target audience, and offer customer-centric promotions. By personalizing campaigns as per customers' preferences, businesses in the hospitality field can significantly enhance the level of engagement with clients and steer the conversion rate (Kim et al. 2025).

In this changing scenario, AI not only opens the door for easy service provision but also allows the formulation of customer-oriented marketing strategies. Analytics solutions developed on the basis of artificial intelligence possess the unprecedented ability to establish different kinds of consumer opinions, social media interaction levels, and online reviews altogether. With such intensive analysis, it is possible to spot emerging trends and areas of improvement. The results derived from the analytical process are of immense worth and are basic tools for companies looking to customize their service offerings and improve their promotional message strategies. Moreover, the results are of immense worth in terms of creating improved customer relationships in the long run (Lee et al. 2021). Nonetheless, artificial intelligence has its disadvantages. Data privacy, automation bias, and the loss of the human touch—a major characteristic of the hospitality industry—remain persistent problems for customers and businesses alike (Huang et al. 2022).

The present research follows a critical analytical approach towards the complex dynamics of AI-powered customer service, consumer sentiment, and online marketing strategies in the hospitality sector, with a focus on the JW Marriott Mumbai Sahar. As a prestigious member of the globally recognized luxury hotels brand Marriott International, the hotel has emerged as a trendsetter in the fast-evolving Indian hospitality sector by embracing AI-powered solutions to enhance customer engagement and overall satisfaction. For example, the hotel has launched a number of path-breaking initiatives, such as AI-based QR code service platforms, sophisticated digital concierge systems, and effective paperless communication systems. These innovations have placed the JW Marriott Mumbai Sahar at the forefront of innovative contactless hospitality solutions, consistently setting new industry standards (Hospitality Lexis 2022).

The application of AI is among global trends. The AI travel and hospitality industry market is projected to grow to USD 1.2 billion by 2026, with a compound annual growth rate (CAGR) of 11.8% from 2022 (Statista 2024). AI-driven chatbots, for instance, are able to answer between 80% of standard guest questions without human support, allowing hotels to redirect human capital into high-touch, emotionally charged service transactions (WorldMetrics 2024a). Besides, over half of global travelers (54%) increasingly prefer to interact with AI-based digital interfaces over human-to-human (WorldMetrics 2024b). The main drivers are efficiency, speed and tailored options.

In the Indian hospitality industry, the JW Marriott Mumbai Sahar experience is a reflection of trends in the making as technology quickly transforms customer expectations. However, amid the rapid growth in digital adoption, culture—in this case, the importance of human interaction in Indian culture—can influence consumer attitudes toward artificial intelligence differently than in Western economies (Deloitte 2023a).

LITERATURE REVIEW

The deployment of Artificial Intelligence (AI) technology in the hospitality industry has transformed customer service, operations, and marketing. Researchers concur that AI-enabled tools—viz. chatbots, virtual assistants, and predictive analytics—enhance operational efficiency, customer interactions, and customization (Limna 2023, Al-Hyari et al. 2023, Kadagidze and Ugrelidze 2023). Nonetheless, the literature also reveals inherent challenges, such as data privacy concerns, astronomically high implementation costs, and the possibility of job displacement (Rasheed et al. 2023, Sardesai et al. 2024). Although AI drives increased guest satisfaction and revenue management, it cannot fully substitute human-centric service, particularly in upscale environments where emotional intelligence and relational service are of paramount importance (Reis 2024). Cultural and regional factors also contribute to AI adoption, including trust, security, and resistance from customers, which affect its adoption in developed and emerging economies (Du 2024, Gajić et al. 2024). AI-based marketing and customer engagement tactics—utilizing tools such as personalization, big data, and green marketing—are established to boost brand trust and customer loyalty; however, ethical implications remain an issue (Ja'afar et al. 2024, Bulchand-Gidumal 2024, Suraña-Sánchez and Aramendia-Muneta 2024). Current research emphasizes the importance of balancing AI with a human touch to attain sustainable and guest-centric service experiences (Mariani and Borghi 2024).

The application of AI in hospitality is transforming operational processes and customer service, fueled by the need for efficiency and personalization (Ivanov and Webster 2019a). Technologies like chatbots, virtual assistants, and data analytics mechanize routine processes and accelerate services, with AI-powered chatbots now responding to up to 80% of customer inquiries (WorldMetrics 2024a). Successful AI adoption, however, relies on consumer behavior, shaped by social influence, ease of use, and generational mindsets—millennials, for instance, are more accepting in F&B environments than in wellness services where human touch is desired (Bhuiyan et al. 2024, Rauf et al. 2022). Integrity of AI systems—accuracy and responsiveness—are determinants of customer satisfaction and brand loyalty (Nguyen et al. 2022). Data privacy, security, and job displacement, however, are complicating issues in the use of AI in hospitality, calling for open practices and ethical foundations (Tussyadiah et al. 2020). Cultural factors also influence the use of AI, with the EU, for instance, having higher trust levels due to strong data protection laws, while Indian guests may still desire human-centric service (Deloitte 2023b). Beyond guest interfaces, AI mechanizes backend processes like inventory management, energy efficiency and predictive maintenance, resulting in cost savings and agility (Casillo et al. 2021).

Complications, however, include high implementation costs, integration issues, staff resistance and risks such as AI-generated fake reviews (AP News 2024, Allam 2020). In the future, innovations such as emotional AI are expected to integrate automation with empathy, necessitating calls for further research on AI-human collaboration and its long-term effect on customer loyalty and competitive edge (Singh et al. 2024, Gupta and Bansal 2023). Doborjeh et al. (2021) systematically reviewed 146 papers to examine AI's dynamic role in hospitality and tourism with a focus on how machine learning and neural networks enhance demand forecasting, customer service, and tourist behavior analysis. Their findings emphasize the significance of personalized AI models in enhancing decision-making and guest experience. Supporting this, Jabeen et al. (2022) investigate AI's revolutionary role through service robots, check-in machines, and virtual assistants that make operations automatic and maximize guest satisfaction, particularly in high-end hotels. Nevertheless, they caution that AI is not yet emotionally smart enough to completely mirror the personalized service required in high-end hospitality facilities. Limna (2023) further investigates AI's central role in contemporary hotel operations, emphasizing its impact on customer service, revenue management, and hyper-personalization. While AI applications enhance efficiency and guest loyalty through predictive analysis and 24/7 support, data privacy, cybersecurity, and the digital divide remain issues.

Based on a bibliometric review of more than 500 studies, Knani et al. (2022) mapped out five clusters of AI in hospitality research: automation, revenue management, personalization, consumer behavior and ethics. Their findings emphasise the importance of AI for operational efficiency and customer satisfaction, as well as growing concerns around data privacy and algorithm transparency. In the same perspective, Huang et al. (2022) proposed a conceptual model for quantifying integration of AI with several dimensions, such as AI Capability, Value Co-Creation, and Impact on service ecosystems. They argue that

AI offers personalized guest journeys at a mass scale and can reshape frontline operations but should be carefully aligned with brand values and people-centric practices in hospitality.

Popesku (2019) explores real-world applications of AI, pointing to tools like Marriott's ChatBotlr and Hilton's Connie, which increase responsiveness in service, facilitate personalization, and optimize operations. Further, Mariani and Borghi (2021) accentuate this subtlety, observing that while AI optimizes efficiency and reduces waiting time, efforts to simulate human interaction may elicit mixed responses from guests. Their sentiment analysis highlights cultural variances in attitudes toward AI, emphasizing that in the paradigms of luxury hospitality, AI must be deployed to supplement—not substitute—human service. Buhalis and Moldavska (2022) highlight the use of AI-powered voice assistants (e.g. Amazon Alexa for Hospitality) in enhancing customer service by offering voice-based, contactless guest engagement, which helps operational productivity and meets a need for contactless interactions following the pandemic.

While the technologies increase convenience and personalization, the paper points out data privacy and surveillance concerns, observing that data policies must be open. In support, Nannelli et al. (2023) classify AI applications as operational automation, CRM, and strategic management, pointing out innovations such as virtual tours and itinerary planners that maximize guest personalization. They also point out the role of AI in sustainability, with smart technologies minimizing environmental effects, while warning against the replacement of employee jobs and the loss of human empathy. Law et al. (2024) also discuss the dominance of service robots in hospitality research, observing their contribution to operational efficiency and customer satisfaction. However, their review highlights gaps in the research, most notably the role of AI in in-stay service delivery and supplier-side issues. They are also seeing a growing interest in generative AI, including a future hospitality trend of using it for marketing and guest communication.

In this regard, Saydam et al. (2022) provide a systematic review on the application of artificial intelligence in the hospitality industry, focusing on the impact of technologies such as service robots, chatbots, and recommendation systems that promote customer satisfaction due to tailored services and improved operational efficiency. Likewise, Bulchand-Gidumal (2022) identifies the possibility of artificial intelligence revolutionizing hospitality, founded on service automation and guest personalization to address the ethical nature of data privacy and job loss. Zahidi et al. (2024) expand upon the dual potential of AI to improve service delivery and backend operations like dynamic pricing and inventory management and warn of challenges, including implementation costs and data security risks. In parallel with these studies, Kim et al. (2025) discuss the evolving role of AI in the areas of technology adoption, operations, and marketing, and provide a stakeholder-centric research agenda to solve managerial and organizational challenges in deploying AI. Finally, Das et al. (2024) examine the impacts of artificial intelligence on hospitality marketing, highlighting how machine learning, natural language processing, and computer vision offer hyper-personalized guest experiences, automation, and data-driven marketing practices. Taken together, these studies reveal the promise of artificial intelligence as a major enabler of customer-centric innovations and operational excellence, while highlighting the need for responsible and sustainable implementation.

METHODOLOGY

This study adopted a quantitative research design with an explanatory research approach to evaluate consumer perceptions of AI-driven customer services within the hospitality industry (Creswell 2009). The research was conducted at JW Marriott, Mumbai, a prominent luxury hotel in India known for its implementation of AI-powered service technologies. The study specifically targeted hotel guests who had experienced first-hand interaction with AI-based solutions such as chatbots, automated concierge services, and voice assistants. A convenience sampling technique was adopted, which is widely used in hospitality research due to its practicality and cost-effectiveness (Saunders et al. 2019). Guests were approached on-site based on their willingness and availability to participate in the survey. A total of 320 responses were approached, out of which 293 valid questionnaires were retained for final analysis after screening them for missing values and incomplete responses. Primary data collection was carried out through a structured questionnaire, administered to the respondents both digitally via tablets and in printed form to ensure maximum participation (Mariani and Borghi 2021). The questionnaire comprised two key sections: the first section gathered demographic information of the respondents, including their age, gender, frequency of hotel stays, etc., while the second section consisted of 14 items measured on a five-point Likert scale for assessing guest responses regarding the experience of AI service attributes such as response time, accuracy, personalization, ease of use, empathy, trust, and overall customer satisfaction. To fulfill the research objectives, a null hypothesis ($H_0 1$) – "There is no significant impact of AI-powered customer service attributes on consumer perception" was formulated and tested by applying Multiple Regression to examine the influence of AI service attributes on the perceptions of hotel guests. This method was chosen for its robustness and effectiveness in identifying the strength and nature of relationships between the variables under investigation (Hair et al. 2019).

ANALYSIS

The collected data was systematically analyzed to study the demographic profile of hotel guests and to determine the impact of AI-powered customer service attributes on consumer perception at JW Marriott, Mumbai. The demographic analysis revealed a diverse pool of hotel guests (n = 293), highlighting the global nature of luxury hotel clientele in India (Popesku 2019). A significant percentage of respondents (n = 175; 59.7%) were aged between 25 and 45 years, representing the group of both business and leisure guests who were typically more inclined towards technology-enabled services (Huang et al. 2022). Gender distribution of the respondents showed 59.7% male (n = 175) and 40.3% female (n = 118) respondents, postulating that male travelers often constitute a larger share of business-related stays in luxury hotels (Mariani and Borghi 2021).

In terms of stay frequency, 45.1% (n = 132) were repeat guests familiar with AI-powered customer service, while 54.9% (n = 161) were first-time visitors, contributing a mix of fresh and seasoned perspectives on AI-enabled service delivery (Knani et al. 2022). Regarding AI familiarity, 67.9% (n = 199) of respondents reported high familiarity with AI-driven services such as chatbots, virtual concierges and voice assistants, whereas 32.1% (n = 94) had limited or no previous exposure (Saydam et al. 2022). Additionally, 70.0% (n = 205) were domestic travelers from various Indian regions, while 30.0% (n = 88) were international guests, thereby enriching the dataset with culturally diverse insights into AI service perceptions (Bulchand-Gidumal 2022). Following the descriptive analysis, a Multiple Linear Regression (MLR) analysis was performed to test the impact of 14 AI-powered customer service attributes on consumer perception, in line with methodologies recommended by Hair et al. (2019) for multi-variable hospitality studies. The independent variables included critical AI service dimensions such as prompt response, minimal delays, accuracy, trust in information, personalized recommendations, customization, ease of use, independence, understanding of concerns, customer appreciation, comfort and reliance, trust in security, enhanced experience, and preference for AI over human representatives (Buhalis and Moldavska 2022, Nannelli et al. 2023).

The dependent variable was consumer perception of AI-powered customer service within the luxury hospitality context. The model summary (Table 11.1) demonstrated a strong relationship between AI service attributes and consumer perception, with an R value of 0.838, indicating a high correlation. The R-squared value of 0.703 suggested that approximately 70.3% of the variance in consumer perception could be explained by the selected AI service attributes, underscoring the model's explanatory power. The adjusted R-squared value of 0.688 further validated the model's robustness by adjusting for potential biases due to multiple predictors (Law et al. 2024).

The standard error of the estimate was 0.20045, indicating a reasonable level of predictive accuracy, while the Durbin-Watson statistic of 1.379 confirmed the absence of significant autocorrelation in the residuals, satisfying one of the key assumptions of regression analysis (Creswell 2014).

The results of ANOVA revealed that the overall regression model was statistically significant ($F = 46.959$, $p < 0.001$), confirming that AI-powered customer service attributes significantly influenced hotel guests' perception at JW Marriott, Mumbai. The total sum of squares was 37.586, with a regression sum of squares of 26.416 and a residual sum of squares of 11.170, emphasizing the regression model's strong explanatory power and validating its suitability for analyzing consumer perception (Hair et al. 2019, Creswell and Creswell 2017).

The coefficients table displayed the relative strength of each independent variable and its impact on hotel guests' perception. The most influential factors were "AI Preference" ($\beta = 0.280$, $p < 0.001$), suggesting that guests who preferred AI-powered customer service over human interaction had a positive and significant perception of AI services, in line with prior findings (Kim et al. 2025, Saydam et al. 2022).

Table 11.1 Individual impact of AI-powered customer service attributes on consumer perception

Model	R	R Square	Adjusted R Square	Std. Error of the Estimate	Durbin-Watson
	.838	**.703**	**.688**	**.20045**	**1.379**
	B	**Std. Error**	**Beta**	**t**	**Sig.**
(Constant)	.497	.177		2.802	.005
Prompt_Response: The AI-powered customer service responds to my queries promptly.	.081	.024	.113	3.360	.001
Minimal Delays: I rarely experience delays when interacting with AI customer service.	.056	.011	.168	4.943	.000
Accuracy: The AI provides accurate and relevant answers to my inquiries.	.083	.036	.123	2.344	.020
Trust Information: I trust the information provided by the AI-powered service.	.015	.037	.021	.393	.695
Personalized Recommendations: The AI recognizes my preferences and provides tailored recommendations.	.081	.031	.117	2.610	.010
Customization: The AI-powered service feels customized to my needs.	.085	.015	.208	5.563	.000
Ease of Use: The AI-powered system is easy to navigate and interact with.	.062	.033	.096	1.860	.064
Independence: I do not need assistance to use the AI-based customer service system.	.086	.034	.146	2.505	.013
Understanding Concerns: The AI-powered service understands my concerns and provides appropriate responses.	.077	.036	.120	2.153	.032
Customer Appreciation: The AI responds in a way that makes me feel valued as a customer.	.037	.031	.051	1.176	.241
Comfort Reliance: I feel comfortable relying on AI-powered customer service for assistance.	.083	.032	.107	2.575	.011
Trust Security: I trust AI to handle my personal information securely.	-.044	.031	-.070	-1.435	.152
Enhanced Experience: AI-powered customer service enhances my overall experience in the hospitality sector.	.078	.033	.108	2.377	.018
AI Preference: I prefer AI-powered customer service over human representatives in certain situations.	.127	.016	.280	7.899	.000
ANOVA Results	**Sum of Squares**	**df**	**Mean Square**	**F**	**Sig.**
Regression	26.416	14	1.887	46.959	.000
Residual	11.170	278	.040		
Total	37.586	292			

"Customization" ($\beta = 0.208$, $p < 0.001$) emerged as another critical driver, indicating that AI services perceived as highly personalized substantially enhanced consumer perception. This supports arguments by Buhalis and Moldavska (2022) and Das et al. (2024), who emphasize the importance of hyper-personalization in shaping positive AI service experiences. Speed-related attributes were also significant: "Minimal Delays" ($\beta = 0.168$, $p < 0.001$) and "Prompt Response" ($\beta = 0.113$, $p = 0.001$) positively influenced consumer perception, reinforcing Nannelli et al. (2023) assertion that efficiency and timeliness are key expectations among modern hotel guests. Further, "Accuracy" ($\beta = 0.123$, $p = 0.020$) and "Personalized Recommendations" ($\beta = 0.117$, $p = 0.010$) contributed significantly, indicating that AI systems capable of providing precise and tailored suggestions improve the guest experience, as also noted by Law et al. (2024) in their review of AI service quality factors. The attribute "Independence" ($\beta = 0.146$, $p = 0.013$) indicated that guests appreciated being able to utilize AI services autonomously, while "Understanding Concerns" ($\beta = 0.120$, $p = 0.032$) and "Comfort Reliance" ($\beta = 0.107$, $p = 0.011$) highlighted the growing importance of AI's ability to create a supportive and seamless customer journey (Zahidi et al. 2024). Finally, "Enhanced Experience" ($\beta = 0.108$, $p = 0.018$) further confirmed AI's positive influence on elevating overall guest satisfaction.

However, several variables did not exhibit a statistically significant impact. "Trust in Information" ($\beta = 0.021$, $p = 0.695$) and "Customer Appreciation" ($\beta = 0.051$, $p = 0.241$) were not significant predictors, indicating that these factors may be less critical in shaping consumer perception of AI-powered services in luxury hospitality settings (Bulchand-Gidumal 2022). "Ease of Use" ($\beta = 0.096$, $p = 0.064$) was marginally significant, suggesting that while intuitive and user-friendly AI interfaces are valued, they may not be as influential as other service dimensions. Notably, "Trust in Security" ($\beta = -0.070$, $p = 0.152$) yielded a negative but non-significant result, implying that, in this context, data security concerns were not a primary determinant of consumer perception, though this could vary depending on guest demographics or AI maturity (Knani et al. 2022). Overall, the study identified AI Preference, Customization, Minimal Delays, Prompt Response, Accuracy, Personalized Recommendations, Independence, Understanding Concerns, and Enhanced Experience as key determinants of consumer perception towards AI-powered customer service in luxury hospitality. These findings suggest that optimizing these critical attributes can significantly enhance customer satisfaction and foster greater acceptance of AI-enabled services in the hospitality sector (Kim et al. 2025, Huang et al. 2022).

CONCLUSIONS

This study concluded that AI-powered customer services are pivotal in molding consumer perceptions within the Indian hospitality sector, particularly in luxury hotels such as JW Marriott, Mumbai. The findings highlighted that key attributes like AI Preference, Customization, Minimal Delays, and Prompt Response strongly influenced the attitude of hotel guests towards AI-driven customer service. The strong predictive regression model confirmed that AI services are well-received by guests when they deliver efficiency, accuracy, and personalized interactions. In contrast, attributes such as Trust in Security and Customer Appreciation were found to be less influential, suggesting that these are considered baseline expectations in luxury hotels in India.

These results support the growing acceptance of AI technologies in commercial hotels, highlighting the critical need for hotel managers to emphasize continuous investment in AI solutions to maintain a competitive edge. The influence of AI Preference, Customization, and Responsiveness demonstrates how AI-driven services bridge the gap between advancements in technology and consumer behavior dynamics.

From a managerial point of view, the study provides implications for hospitality stakeholders. Hotel managers are encouraged to enhance service quality by focusing on personalization, reducing service latency, and providing seamless and responsive interactions between guests and AI-powered systems. At the same time, technology developers should refine AI-driven tools such as chatbots, virtual assistants, and recommendation engines to better position themselves with evolving guest expectations through smart, real-time, and adaptive delivery of service.

Furthermore, decision-makers have an important role to play in this scenario by cultivating a regulatory environment that adheres to responsible AI implementation, ensuring guest data privacy, and ethical usage of AI in hospitality sectors. Creating detailed guidelines will help balance technological advancement with consumer trust and well-being. Ultimately, this study emphasizes the transformative power of AI within the hospitality context, focusing on continuous innovation and guest-centric design that are key to achieving customer satisfaction and loyalty. By effectively handling AI-powered customer interactions, luxury hospitality stakeholders can strengthen their brand positioning and remain competitive in a dynamic service landscape. In conclusion, the findings highlight that AI is no longer an auxiliary tool but a core strategic suit that can revamp service delivery models to meet the dynamic customer expectations in the modern world.

LIMITATIONS AND STUDY FORWARD

The findings of the study are restricted to its focus on a single luxury hotel, JW Marriott Mumbai, which may not be truly representative of other hotel categories. Adopting a convenience sampling technique may introduce respondent bias and may impact sample representativeness. Additionally, the study only captures consumer perceptions, excluding managerial viewpoints and operational challenges. Lastly, self-reported survey data may be subject to response bias.

Future research should include diverse hotel categories and regions to improve generalizability. Using probability sampling could reduce bias and enhance representativeness. Incorporating qualitative methods like interviews with hotel staff can offer insights into AI adoption challenges. Longitudinal studies tracking changing consumer perceptions and AI sentiment analysis of guest reviews could further enrich findings.

CONFLICT OF INTEREST AND ETHICAL STANDARDS

This study adheres to the highest ethical standards in research and publication. The authors declare no conflict of interest that could influence the design, methodology, analysis, or findings of this research. All data collected and analyzed for this study have been obtained with appropriate permissions and in compliance with ethical guidelines.

Participants in this study were fully informed about the purpose of the research, and their consent was obtained before data collection. Confidentiality and anonymity were strictly maintained to ensure the privacy of respondents.

Furthermore, AI-driven customer service and its implications on consumer perception were analyzed objectively, without bias, ensuring that the study contributes meaningfully to academic and industry discourse. Any potential ethical concerns related to AI adoption in hospitality, such as data privacy and algorithmic bias, have been duly acknowledged and addressed within the scope of this research.

ACKNOWLEDGEMENT

I would like to express my sincere gratitude to all individuals and institutions that contributed to the successful completion of this research study.

Firstly, I extend my appreciation to GNA University, Sanskaram University for providing the necessary support and resources to conduct this study.

I am also grateful to the management and staff of **JW Marriott Mumbai** for their cooperation in facilitating data collection and providing valuable insights into AI-driven customer service practices. Their assistance has been instrumental in shaping the findings of this research.

Special thanks to the participants who generously shared their experiences and perceptions, making this study possible. Their honest responses have provided valuable perspectives on the impact of AI in the hospitality industry.

I would also like to acknowledge the guidance and encouragement of my mentors, colleagues and peers, whose feedback has been invaluable in refining this research. Lastly, I am deeply thankful to my family and friends for their unwavering support and motivation throughout this academic endeavor.

AUTHOR'S CONTRIBUTION

Ishan Bakshi: Data Analysis, Manuscript Writing
Hitesh Marwaha: Methodology, Data Collection, Validation
Aashish Samuel: Conceptualization, Review amd Editing
Ashish Ahlawat: Literature Review, Final Manuscript Review, Corresponding Author

REFERENCES

Al-Hyari, H.S., Al-Smadi, H.M. and Weshah, S.R. 2023. The impact of artificial intelligence (AI) on guest satisfaction in hotel management: An empirical study of luxury hotels. GeoJournal of Tourism and Geosites. 48(2 supplement): 810–819. https://doi.org/10.30892/gtg.482spl15-1081

Allam, Z. 2020. The rise of machine intelligence: Digital governance, smart cities and risk society. Springer.

AP News. 2024. The internet is rife with fake reviews. Will AI make it worse? Retrieved from https://apnews.com/article/40f5000346b1894a778434ba295a0496apnews.com

Bhuiyan, K.H., Ahmed, S. and Jahan, I. 2024. Consumer attitude toward using artificial intelligence (AI) devices in hospitality services. Journal of Hospitality and Tourism Insights. 7(2): 968–985. https://doi.org/10.1108/JHTI-08-2023-0551Emerald

Buhalis, D. and Moldavska, I. 2022. Voice assistants in hospitality: Using artificial intelligence for customer service. Journal of Hospitality and Tourism Technology. 13(3): 473–487. https://doi.org/10.1108/JHTT-03-2021-0074

Bulchand-Gidumal, J. 2022. Impact of artificial intelligence in travel, tourism and hospitality. *In*: Gretzel, U., Fuchs M. and Zanker, M. (eds). Handbook of e-Tourism. 1–25. Springer. https://doi.org/10.1007/978-3-030-48652-5_110

Bulchand-Gidumal, J. 2024. Artificial intelligence's impact on hospitality and tourism marketing: Exploring key themes and addressing challenges. (Working paper/preprint PDF). Bournemouth University Repository.

Casillo, M., De Pietro, O. and Gallo, L. 2021. AI-based systems for personalized recommendations in the tourism sector. Information. 12(3): 124.

Creswell, J.W. 2009. Research Designs. Qualitative, Quantitative, and Mixed Methods Approaches, Third Edition. SAGE Publications.

Creswell, J.W. and Creswell, J.D. 2017. Research Design: Qualitative, Quantitative, and Mixed Methods Approaches, Fifth Edition. SAGE Publications.

Das, I.R., Talukder, M.B. and Kumar, S. 2024. Implication of artificial intelligence in hospitality marketing. *In*: Kumar, S., Nayak, A. and Narayan, V. (eds). Utilizing Smart Technology and AI in Hybrid Tourism and Hospitality. 209–221. IGI Global. https://doi.org/10.4018/978-1-6684-9863-1.ch012

Deloitte. 2023a. 2023 India digital consumer trends survey. Deloitte Insights. Retrieved March 24, 2025, from https://www2.deloitte.com/in/en/pages/technology-media-and-telecommunications/articles/digital-consumer-trends-india.html

Deloitte. 2023b. Digital Consumer Trends Survey 2023: India. Deloitte Insights. Retrieved from https://www2.deloitte.com

Doborjeh, Z., Hemmington, N., Doborjeh, M. and Kasabov, N. 2021. Artificial intelligence: A systematic review of methods and applications in hospitality and tourism. International Journal of Contemporary Hospitality Management. 34(3): 1154–1176. https://doi.org/10.1108/IJCHM-06-2021-0767

Du, J. 2024. Do humans trust AI in HRM? Why do? Why not?—Insights from a decade of research. Journal of Research in Social Science and Humanities. 3(7): 20–48.

Hair, J.F., Black, W.C., Babin, B.J., Anderson, R.E. and Tatham, R.L. 2019. Multivariate Data Analysis, Eighth Edition. Cengage Learning, EMEA

Hospitality Lexis. 2022. New technology and digitalization initiatives at JW Marriott Mumbai Sahar. Hospitality Lexis. Retrieved March 24, 2025, from https://

hospitalitylexis.media/new-technology-and-digitalization-initiatives-at-jw-marriott-mumbai-sahar/

Huang, A., Chao, Y. and de la Mora Velasco, E. 2022. When artificial intelligence meets the hospitality and tourism industry: An assessment framework to inform theory and management. Journal of Hospitality and Tourism Management. 51: 457–468. https://doi.org/10.1016/j.jhtm.2022.03.004

Gajić, T., Vukolić, D., Bugarčić, J., Đoković, F., Spasojević, A., Knežević, S., et al. 2024. The adoption of artificial intelligence in Serbian hospitality: A potential path to sustainable practice. Sustainability. 16(8): 3172. https://doi.org/10.3390/su16083172

Gupta, T. and Bansal, S. 2023. AI-Driven emotional recognition in digital ads: A novel approach to consumer engagement. Journal of Marketing & Supply Chain Management. 2(3): 1–7. https://doi.org/10.11648/j.ajcst.20240704.13

Ivanov, S. and Webster, C. 2019a. Conceptual framework of the use of robots, artificial intelligence and service automation in travel, tourism and hospitality companies. International Journal of Contemporary Hospitality Management. 31(2): 155–172. https://doi.org/10.1108/IJCHM-09-2017-0579

Ivanov, S. and Webster, C. 2019b. Robots, Artificial Intelligence and Service Automation in Travel, Tourism and Hospitality. Emerald Group Publishing.

Ja'afar, N.S., Khan, N., Oh Kim Seng, V. and Obreja, S.G. 2024. AI-Powered digital marketing: Elevating brand perception in the event industry. Journal of Advanced Research in Applied Sciences and Engineering Technology. 64(1): 75–94.

Jabeen, F., Al Zaidi, S. and Al Dhaheri, M.H. 2022. Automation and artificial intelligence in hospitality and tourism. Tourism Review. 77(2): 533–548. https://doi.org/10.1108/TR-07-2021-0341

Kadagidze, L. and Ugrelidze, E. 2023. AI in hospitality industry: A comprehensive study on its impact on operations, customer experience, and revenue management. *In*: Proceedings of the VII International European Conference on Interdisciplinary Scientific Research.

Kim, H., So, K.K.F., Shin, S. and Li, J. 2025. Artificial intelligence in hospitality and tourism: Insights from industry practices, research literature and expert opinions. Journal of Hospitality and Tourism Research. Advance online publication. https://doi.org/10.1177/1096348024123456

Knani, M., Echchakoui, S. and Ladha, R. 2022. Artificial intelligence in tourism and hospitality: Bibliometric analysis and research agenda. International Journal of Hospitality Management. 103: 103217. https://doi.org/10.1016/j.ijhm.2022.103217

Law, R., Lin, K.J., Ye, H. and Fong, D.K.C. 2024. Artificial intelligence research in hospitality: A state-of-the-art review and future directions. International Journal of Contemporary Hospitality Management. 36(6): 2049–2068. https://doi.org/10.1108/IJCHM-02-2023-0189

Lee, M., Kwon, W. and Back, K.J. 2021. Artificial intelligence for hospitality big data analytics: Developing a prediction model of restaurant review helpfulness for customer decision-making. International Journal of Contemporary Hospitality Management. 33(7): 2362–2383. https://doi.org/10.1108/IJCHM-10-2020-1203

Limna, P. 2023. Artificial Intelligence (AI) in the hospitality industry: A review article. International Journal of Computer Applications. 182(9): 12–19. https://doi.org/10.5120/ijca2023923456

Mariani, M. and Borghi, M. 2021. Customers' evaluation of mechanical artificial intelligence in hospitality services: A study using online reviews analytics. International Journal of Contemporary Hospitality Management. 33(11): 4061–4082. https://doi.org/10.1108/IJCHM-07-2020-0772

Marriott International. 2023. Serve360: Doing good in every direction—2023 Sustainability and Social Impact Report. Retrieved from https://serve360.marriott.com

Nannelli, M., Capone, F. and Lazzeretti, L. 2023. Artificial intelligence in hospitality and tourism. State of the art and future research avenues. European Planning Studies. 31(3): 537–555. https://doi.org/10.1080/09654313.2022.2132512

Nguyen, T.-M., Quach, S. and Thaichon, P. 2022. The effect of AI quality on customer experience and brand relationship. Journal of Consumer Behaviour. 21(3): 481–493. https://doi.org/10.1002/cb.1974

Popesku, J. 2019. Current applications of artificial intelligence in tourism and hospitality. *In*: Sinteza 2019—International Scientific Conference on ICT and E-Business Related Research. 283–290. Singidunum University. https://doi.org/10.15308/Sinteza- 2019-283-290

Rasheed, H.M.W., He, Y., Khizar, H.M.U. and Abbas, H.S.M. 2023. Exploring consumer-robot interaction in the hospitality sector: Unpacking the reasons for adoption (or resistance) to artificial intelligence. Technological Forecasting and Social Change. 192: 122555. https://doi.org/10.1016/j.techfore.2023.122555

Rauf, A., Zurcher, M., Pantelidis, I.S. and Winbladh, J. 2022. Millennials' perceptions of artificial intelligence in hotel service encounters. Consumer Behavior in Tourism and Hospitality. 17(1): 3–16. https://doi.org/10.1108/CBTH-04-2021-0104

Reis, J. 2024. Customer service through AI-powered human–robot relationships: Where are we now? The case of Henn na Cafe, Japan. Technology in Society. 77: 102570. https://doi.org/10.1016/j.techsoc.2024.102570

Sardesai, S., D'Souza, E. and Govekar, S. 2024. Analysing the impacts of artificial intelligence service quality and human service quality on customer satisfaction and customer loyalty in the hospitality sector. Turizam. 28(1): 2–211. https://doi.org/10.5937/turizam28-45450

Saunders, M.N., Lewis, P. and Thornhill, A. 2019. Research Methods for Business Students (Eighth). Harlow: Pearson Education Limited.

Saydam, M.B., Arici, H.E. and Koseoglu, M.A. 2022. How does the tourism and hospitality industry use artificial intelligence? A review of empirical studies and future research agenda. Journal of Hospitality Marketing and Management. 31(8): 908–936.

Singh, A.P., Saxena, R. and Saxena, S. 2024. The human touch in the age of artificial intelligence: A literature review on the interplay of emotional intelligence and AI. Asian Journal of Current Research. 9(4): 36–50. https://doi.org/10.56557/ajocr/2024/v9i48860

Statista. 2024. Artificial Intelligence in the Travel and Hospitality Industry—Worldwide. Statista. Retrieved March 24, 2025, from https://www.statista.com/statistics/ai-hospitality-market-size/

Suraña-Sánchez, C. and Aramendía-Muneta, M.E. 2024. Impact of artificial intelligence on customer engagement and advertising engagement: A review and future research agenda. International Journal of Consumer Studies. 48(2): e13027. https://doi.org/10.1111/ijcs.13027.

Tussyadiah, I., Park, S. and Fesenmaier, D.R. 2020. Assessing the effectiveness of AI-powered customer service: A customer perspective. International Journal of Hospitality Management. 89(102528): 1–12. https://doi.org/10.1016/j.ijhm.2020.102528

WorldMetrics. 2024a. Chatbots in hospitality industry statistics: Market data report 2024. WorldMetrics. Retrieved March 24, 2025, from https://www.worldmetrics.org/chatbots-in-hospitality-industry-statistics/

WorldMetrics. 2024b. AI in the hospitality industry statistics: Market data report 2024. WorldMetrics. Retrieved March 24, 2025, from https://worldmetrics.org/ai-in-the-hospitality-industry-statistics/

Zahidi, F., Kaluvilla, B.B. and Mulla, T. 2024. Embracing the New Era: Artificial Intelligence and Its Multifaceted Impact on the Hospitality Industry. Journal of Open Innovation: Technology, Market and Complexity. 10(1): 100390. https://doi.org/10.1016/j.joitmc.2024.100390

Chapter **12**

Enhancing Pre-flight Preparedness through Virtual Airport Tours: A User-Centric Study of Indian Air Passengers

Uma Pandey*[1], Parikshit Sharma[2], Isha Pabia[3] and Avinash Kumar Chauhan[4]

[1]Assistant Professor, Jagran School of Hospitality and Tourism, Jagran Lakecity University, Bhopal, M.P., India. Orcid id: 0000-0002-8817-1546

[2]Associate Professor, Jagran School of Hospitality and Tourism, Jagran Lakecity University, Bhopal, M.P., India. Orcid id: 0000-0002-2231-0865

[3]Research Scholar, Jagran School of Hospitality and Tourism, Jagran Lakecity University, Bhopal, M.P., India.

[4]Research Scholar, School of Management and Entrepreneurship, IT Jodhpur, Rajasthan, India.

INTRODUCTION

The exponential growth of technology has changed all aspects of human life, including travel planning procedures and experiences, as well as how they reflect on travel (Pencarelli 2019). Technological advances have remained vital elements in aviation since they continuously improve safety performance while increasing operational efficiency alongside delivering superior customer service. Passenger

*For Correspondence: umatripathi86@gmail.com

DOI: 10.1201/9781003540083-12

experience enrichment through Virtual Reality (VR) and Augmented Reality (AR) (McGill et al. 2022) has gained increasing attention during the recent period. The concept of virtual airport tours links to digitally simulated walkthroughs, which render airport terminals and processes, allowing passengers to learn the environment before they physically arrive there. An academic investigation should focus on this underutilised area because such technological solutions have shown promise in different regions around the world, but their implementation and adoption in India remain limited. Air travel recovery from the global pandemic presents dual challenges for airports and airlines by requiring safety standards as well as clean and stress-free travel services for passengers. First-time flyers, elderly passengers, and people with accessibility issues, as well as those with high levels of travel anxiety, experience overwhelming feelings in airports. Travellers face problems during unfamiliar airport check-in stages, security procedures, boarding procedures and terminal route navigation, which results in both confusion and delayed journeys as well as negative impressions (Agrawal et al. 2024). The combination of virtual airport tours thus functions as an effective pre-flight preparation method to minimize traveller uncertainty and significantly raise their level of preparedness.

The Technology Acceptance Model (TAM) is known as a powerful tool to learn the adoption of different technologies by users. It implies that users' adoption of any new system depends to a great extent on two factors: perceived usefulness (how useful the technology is in performing better) and perceived ease of use (how easy it is to use) (Abdullah et al. 2016). This model has been extensively utilised in research examining digital tools and services across diverse industries, including aviation and tourism. Utilising TAM to understand virtual airport tours enables us to explore how perceptions of usefulness and ease of use among travellers determine their intentions to use these tools during pre-flight preparation. Virtual airport tours deliver interactive media content to public viewers that demonstrates how facilities operate and their procedures. Web applications and mobile applications pair up with VR headsets to present simulative multimedia content that includes progressive building explanations, direction-based services and immigration personalisation programmes. These digital interfaces serve both IT-intelligent passengers and inexperienced passengers by offering them preparation tools, thus reducing pre-travel stress and increasing the level of air travel confidence. Several advantages emerge as passengers reach their transport destination with better knowledge, reduced anxiety and increased satisfaction levels.

The adoption of virtual airport tours within the Indian aviation system has not been explored. India stands as one of the top global markets for air travel expansion because it sees increasing numbers of first-time flyers joining the aviation network annually (Samunderu 2024). Statistics from the Directorate General of Civil Aviation (DGCA) demonstrate that India records an annual passenger number increase, which shows particular growth in the airports serving tier-2 and tier-3 regions. The inexperienced nature of these new air passengers with digital tools and international airports makes them ideal for virtual orientation education. Indian passengers possess varying technological understanding because of cultural elements combined with their digital skill levels and technological exposure experience.

No experimental studies exist about how Indian passengers interpret virtual airport tours, as it is not so popular in the nation. The studies on international airports focus on digital system implementation methods and evaluate Western aviation facilities in addition to these implementation methods. Studies about these tools remain underdeveloped because Indian economy passengers and their behavioural patterns have not been studied through empirical studies. The success of implementation depends on understanding how users perceive these tools since adoptive measures and solution designs need to fit users' perceptions (Fakhr Hosseini et al. 2022). Changes in Indian air traveller demographics constitute an important factor within this context. The development of affordable flight options has enabled numerous people across different economic levels to experience air travel for the first time. Affordable air travel has enabled diverse segments of society to fly, yet it requires new adjustments regarding passenger service standards, airport directions and passenger information distribution. Anxious and inexperienced passengers do not always find traditional static signage, brochures and staff assistance sufficient to handle their concerns (Peterson et al. 2022). As a complementary tool, virtual tours supply standardised yet personalised pre-travel information at the same time. Public expectations relating to travel underwent drastic changes because of the COVID-19 pandemic. Travellers globally have made hygiene and safety measures, together with contactless procedures and mental preparation, the top priorities for their journeys. Travellers in the modern experience may require the preview and mental rehearsal components of a virtual airport tour to become essential elements of a safe travel field. Alongside being a health-conscious practice, this technology supports passenger control while also assuring them that their needs are being met.

Virtual airport tour applications do exist in the Indian market, although not at full scale. A number of private airports currently operate digital map systems as well as guided video tours and 360-degree camera views (Hemmings et al. 2023). These current virtual airport tour initiatives face two main drawbacks: reduced scope of action and poor integration, together with limited ability to engage users throughout the passenger journey. Such tools receive little public awareness, especially in non-metropolitan regions of India. Research needs to determine what knowledge Indian passengers possess about virtual airport tours, along with their expectations and preferred features. There is a knowledge gap between the perception and expectations of Indian air passengers towards virtual airport tours. This study aims to explore and fill this knowledge gap towards virtual airport tours. A virtual tour provided in English creates a barrier for many Indian users since they prefer to use their native regional languages. Users who lack digital experience may avoid systems that provide interfaces with complex designs. This research strives to reveal detailed preferences of users through its efforts to address the mentioned issues.

This study investigates the thoughts and anticipated views of Indian passengers regarding their use of virtual airport visitation tools. The study evaluates passengers' understanding of such tools and their willingness to adopt them while exploring the elements that affect tool perception. This study uses user-focused research to deliver practical recommendations that help airport management teams and airline companies, along with technology developers who seek to deploy or advance virtual orientation systems in Indian airports. Virtual airport tours hold significant potential as an understated enhancement method for customer travel experiences, particularly within the dynamic and multifield Indian market (Heiets et al. 2022, Jung et al. 2017). An analysis of customer views about their preferences allows researchers to detect barriers to implementation and discover methods for improvement. The fundamental requirements of every travel stage must include barrier-free digital interfaces with localised online experiences since these tools enable passengers' success. The achievement of superior service excellence in Indian aviation services alongside passenger engagement success depends extensively on the acquisition of such information.

LITERATURE REVIEW

The aviation industry has regularly recognised the potential of digital technologies to improve the experiences of passengers, like biometric check-in systems, digital boarding passes, airport navigation apps, etc. (Şahin 2024). To make passengers aware of the airport environment, a few airports have introduced the concept of virtual airport tours, which is intended to benefit not only first-time travellers but also those who have travel anxiety. Virtual airport tours are a great way of pre-flight familiarisation. They provide a simulated walkthrough of the airport environment, where passengers can visualise procedures, locate facilities, and prepare themselves for the further journey mentally. In India, as this facility is not widely available, it is necessary to understand the perception and expectations of Indians about virtual airport tours. In order to study this, we are analysing the relationship between three variables, i.e. Perceived Usefulness (PU), Perceived Preparedness (PP), and Intended Use (IU).

Perceived Usefulness and Perceived Preparedness

According to Davis (1989), perceived usefulness is the degree to which an individual believes that using a particular system would enhance their performance in any allocated task. Passengers explore destinations like cities, museums, historical sites, etc. online to get an idea about them and to make navigation easier when they actually reach there (Pencarelli 2019). At airports, virtual tours can be used to create a smoother airport experience by avoiding confusion about airport processes. Many researchers have found that perceived usefulness influences the readiness and confidence of individuals (Joshi et al. 2021, Gestiardi et al. 2021, Tahar et al. 2020). This suggests that when passengers find virtual airport tours useful, they are more likely to feel prepared, both mentally and logistically, for their journey. Thus, we are proposing the below hypothesis:

H1: Perceived usefulness of virtual airport tours positively influences passengers' perceived preparedness for air travel.

method and was conducted online, which may have resulted in a sample skewed towards more digitally literate individuals. As a result, the voices of those who are less familiar with or have limited access to digital tools, particularly older adults, rural travellers, or individuals from lower socio-economic backgrounds, may not be adequately represented. Future research should aim for a more inclusive and representative sampling strategy, perhaps through field surveys at airports or community-based outreach, to ensure a broader demographic perspective. Second, the study's cross-sectional design captures perceptions at a single point in time and does not reflect how these attitudes may change after actual use of a virtual tour. Longitudinal or experimental studies could offer deeper insight into behavioural shifts, user retention, and the long-term effectiveness of virtual airport tours. Incorporating pre- and post-exposure assessments could help establish causal relationships between exposure to virtual tours and changes in travel confidence or satisfaction. Moreover, the research focused primarily on three constructs: Perceived Usefulness, Perceived Preparedness, and Intended Use, which, while meaningful, do not fully capture the broader psychological, technological, and contextual factors that may influence adoption. Future studies should consider expanding the conceptual framework to include variables such as digital trust, perceived ease of use, language preferences, and accessibility concerns, which emerged as important themes in the qualitative responses. Additionally, the study did not investigate how language barriers or regional cultural preferences might shape the acceptance of virtual tours. Since India is a multilingual society with diverse traveller needs, future research should explore how localisation through language options, culturally relevant interfaces, or region-specific airport simulations affects user engagement. Lastly, while many respondents expressed a preference for mobile applications, this study did not explore the technical design aspects that might enhance usability or encourage continued engagement. Future collaborations with interface designers, airport authorities, and behavioural scientists can help develop prototypes that are not only functional but also inclusive and emotionally supportive. In addressing these limitations, future research and implementation efforts can help refine virtual airport tours into a powerful tool that supports India's evolving air travel landscape, ensuring it is accessible, adaptive, and genuinely useful for all segments of society.

REFERENCES

Abdous, M. 2019. Influence of satisfaction and preparedness on online students' feelings of anxiety. The Internet and Higher Education. 41: 34–44. https://doi.org/10.1016/j.iheduc.2019.01.001

Abdullah, F., Ward, R. and Ahmed, E. 2016. Investigating the influence of the most commonly used external variables of TAM on students' Perceived Ease of Use (PEOU) and Perceived Usefulness (PU) of e-portfolios. Computers in Human Behavior. 63: 75–90. https://doi.org/10.1016/j.chb.2016.05.014

Agrawal, S., Bansal, S. and Kamra, V. 2024. Navigating the Skies: A Comprehensive Analysis of Airport Automation and Its Impact on Passenger Experience. *In*: IEEE. 1–6. https://doi.org/10.1109/iscs61804.2024.10581214

Carroll, P.J. 2013. Preparedness. *In*: Handbook of the Uncertain Self. 266–290. Psychology Press.

Davis, F.D. 1989. Perceived usefulness, perceived ease of use, and user acceptance of information technology. MIS Quarterly. 13(3): 319. https://doi.org/10.2307/249008

Fakhr Hosseini, S., Chan, K., Lee, C., Jeon, M., Son, H., Rudnik, J., et al. 2022. User Adoption of Intelligent Environments: A review of technology adoption models, challenges, and prospects. International Journal of Human-Computer Interaction. 40(4): 986–998. https://doi.org/10.1080/10447318.2022.2118851

Fan, X., Jiang, X. and Deng, N. 2022. Immersive technology: A meta-analysis of augmented/virtual reality applications and their impact on tourism experience. Tourism Management. 91: 104534. https://doi.org/10.1016/j.tourman.2022.104534

Fennell, D.A. 2017. Towards a model of travel fear. Annals of Tourism Research. 66: 140–150. https://doi.org/10.1016/j.annals.2017.07.015

Gestiardi, R., Sarwanto, S., Chumdari, C. and Maryani, I. 2021. Using an technology readiness model to understand perceived usefulness of learning in the COVID-19 era. International Journal of Elementary Education. 5(4): 631. https://doi.org/10.23887/ijee.v5i4.39188

Heiets, I., La, J., Zhou, W., Xu, S., Wang, X. and Xu, Y. 2022. Digital transformation of airline industry. Research in Transportation Economics. 92. https://doi.org/10.1016/j.retrec.2022.101186

Hemmings, P.J., Mulheron, M., Murphy, R.J. and Prescott, M. 2023. Innovation in Airport Infrastructure and the Transition to Net Zero.

Jang, J., Ko, Y., Shin, W.S. and Han, I. 2021. Augmented Reality and Virtual Reality for Learning: An examination using an Extended Technology acceptance model. IEEE Access. 9: 6798–6809. https://doi.org/10.1109/access.2020.3048708

Jegundo, A.L., Dantas, C., Quintas, J., Dutra, J., Almeida, A.L., Caravau, H., et al. 2020. Perceived usefulness, satisfaction, ease of use and potential of a virtual companion to support the care provision for older adults. Technologies. 8(3): 42. https://doi.org/10.3390/technologies8030042

Joshi, S., Sharma, M., Bisht, P. and Singh, S. 2021. Explaining the factors influencing consumer perception, adoption readiness, and perceived usefulness toward digital transactions: Online retailing experience of millennials in India. Journal of Operations and Strategic Planning. 4(2): 202–223. https://doi.org/10.1177/2516600x211052649

Jung, T., Dieck, M.C.T., Moorhouse, N. and Dieck, D.T. 2017. Tourists' experience of Virtual Reality applications. 2017 IEEE International Conference on Consumer Electronics, ICCE 2017. 208–210. https://doi.org/10.1109/ICCE.2017.7889287

Kim, K., Hwang, J., Zo, H. and Lee, H. 2014. Understanding users' continuance intention toward smartphone augmented reality applications. Information Development. 32(2): 161–174. https://doi.org/10.1177/0266666914535119

McGill, M., Li, G., Ng, A., Bajorunaite, L., Williamson, J., Pollick, F., et al. 2022. Augmented, virtual and mixed reality passenger experiences. *In*: Studies in Computational Intelligence. 445–475. https://doi.org/10.1007/978-3-030-77726-5_17

Neal, J.D., Sirgy, M. and Uysal, M. 1999. The role of satisfaction with Leisure Travel/ Tourism Services and experience in satisfaction with Leisure Life and Overall Life. Journal of Business Research. 44(3): 153–163. https://doi.org/10.1016/s0148-2963(97)00197-5

Pencarelli, T. 2019. The digital revolution in the travel and tourism industry. Information Technology and Tourism. 22(3): 455–476. https://doi.org/10.1007/s40558-019-00160-3

Peterson, C.M., Birkeland, R.W., Barsel, S., Statz, T.L., Gaugler, J.E. and Finlay, J.M. 2022. Sick with stress: Perspectives on airport travel from persons living with dementia and their travel companions. Disability and Society. 39(1): 234–254. https://doi.org/10.1080/09687599.2022.2070060

Şahin, G.G. 2024. The synergy between tourism, aviation, and digital transformation. *In*: Advances in Mechatronics and Mechanical Engineering (AMME) Book Series. 126–149. https://doi.org/10.4018/979-8-3693-0732-8.ch007

Samunderu, E. 2024. The scope of the global aviation industry. *In*: Advances in African Economic, Social and Political Development. 1–116. https://doi.org/10.1007/978-3-031-61864-2_1

Shore, L., Power, V., De Eyto, A. and O'Sullivan, L. 2018. Technology acceptance and user-centred design of assistive exoskeletons for older adults: A commentary. Robotics. 7(1): 3. https://doi.org/10.3390/robotics7010003

Tahar, A., Riyadh, H.A., Sofyani, H. and Purnomo, W.E. 2020. Perceived ease of use, perceived usefulness, perceived security and intention to use E-Filing: the role of Technology readiness. Journal of Asian Finance Economics and Business. 7(9): 537–547. https://doi.org/10.13106/jafeb.2020.vol7.no9.537

Venkatesh, N., Thong, N. and Xu, N. 2012. Consumer Acceptance and use of Information technology: Extending the unified theory of acceptance and use of technology. MIS Quarterly. 36(1): 157. https://doi.org/10.2307/41410412

Wong, K.T., Osman, R.B., Goh, P.S.C. and Rahmat, M.K. 2013. Understanding student teachers' behavioural intention to use technology: Technology acceptance model (TAM) validation and testing. DOAJ: Directory of Open Access Journals. https://doaj.org/article/00a863d2329d4c7fabac61e0564c7465

Chapter 13

Role of Human-AI Co-Creation in Personalized Itinerary Design and Customer Satisfaction in Smart Tourism

Rahul Singh Shekhawat*[1] and Udaidip Singh Chauhan[2]

[1]Assistant Professor, School of Business, Mody University

[2]Assistant Professor, Faculty Hotel Management and Catering Technology, Parul University

INTRODUCTION

The modern tourism industry is radically altered by Artificial Intelligence (AI) technologies as they rapidly advance and integrate into the industry (Buhalis and Leung 2018). Personalized systems have emerged to assist in the development of travel itineraries, among other areas involving AI usage in personalizing the travel experience (Ricci et al. 2011). Such systems are designed to provide travelers with automated planning and offer pre-made itineraries according to their individual preferences. However, the most pressing question regarding these services is the level of satisfaction achieved among customers by such innovations.

In this developing world, early research on AI in tourism was generally focused on automation and efficiency. In contrast, growing evidence suggests that the factors leading to user satisfaction are not related exclusively to technological

*For Correspondence: rahulsinghshekhawat.sob@modyuniversity.ac.in, udaidip.chauhan29417@paruluniversity.ac.in

DOI: 10.1201/9781003540083-13

capabilities (Maadi, et al. 2021). The nature of interaction and the human level of collaboration between travelers and AI systems are gaining increasing importance. This paradigm shift is called Human-AI Co-Creation, which represents an active partnership whereby tourists work with AI tools in developing particular travel plans that inspire ownership and potentially lead to richer and more relevant experiences (Dellermann et al. 2019).

As Human-AI Co-Creation increasingly shapes personalized travel experiences, its effect on customer satisfaction has not yet been studied in-depth within the smart tourism context (Takuli et al. 2024). Most studies on AI in tourism have addressed technology acceptance, perceived usefulness, and trust regarding AI (Davis 1989, Venkatesh et al. 2003). However, the role of co-creation in customer satisfaction outcomes has not been well explored. The tourism context continues to grow digitized and transforms hyper-personalization expectations among travelers. Therefore, inquiry is required regarding whether co-designing travel purposely with AI tools increases consumers' satisfaction with the trip as a whole in addition to how they planned it.

LITERATURE REVIEW

Human-AI Co-Creation and Customer Satisfaction in Smart Tourism

Smart tourism is the new realm where AI technology effectively enhances experiences, from accessing information to personalized services during travel (Buhalis and Leung 2018). Creating personalized travel itineraries is a major application of AI, which aims to be efficient and tailored for tourist needs (Ricci et al. 2011). However, simply introducing AI does not guarantee customer satisfaction. Emerging research indicates that it is the nature of interaction between the user and the AI systems—the level of collaborative involvement—that becomes most critical in understanding how perceptions and satisfaction outcomes are constructed (Luger and Sellen 2016).

This literature review brings together existing discussions on AI in tourism and personalization and the newly coined idea of Human-AI Co-Creation, highlighting the main gap that this study addresses.

AI and Personalization in Tourism

Artificial Intelligence being incorporated into travel makes it possible to go beyond personalization and aims to hyper-personalize services and recommendations to tailor very specific needs to individual travelers (Gretzel et al. 2015). AI-driven recommendation systems analyze huge datasets of customer behaviors, their preferences, and contextual information to suggest relevant destinations, activities, accommodations, and means of transportation (Fesenmaier et al. 2003). These systems aim to enhance efficiency, save time for travelers, and offer the best travel experience. A number of studies have examined the effectiveness of various

AI techniques such as collaborative filtering, content-based filtering, and hybrid approaches in personalized recommendations (Ricci et al. 2011).

Early research has sometimes focused more on technology than on the user's role in personalization. While personalization aims to meet individual needs, the black-box nature of some AI algorithms often prevents users from understanding their workings, leading to issues with trust and transparency (Tintarev and Masthoff 2015). The user's role becomes crucial because the output largely depends on their input and interaction.

Customer Satisfaction in Smart Tourism

Customer satisfaction is a core success factor in tourism industry development, influencing repeat visitation, positive word-of-mouth, and overall business performance (Oliver 1980). Satisfaction in smart tourism concerns different variables, including information quality, service efficiency, personalization, and technological experiences (Wei et al. 2024). Related research has concentrated on the influence of technology adoption on tourist satisfaction, highlighting perceived usefulness, ease of use, and trust in technology (Davis 1989, Venkatesh et al. 2003).

AI-powered tools provide a wide range of benefits, but they also present problems that might reduce satisfaction, such as algorithmic bias, lack of human touch, and cases where users feel dominated by technology (Cheng et al. 2020). Hence, it is of utmost importance to examine how to design AI systems to enhance user experience and increase satisfaction.

Emergence of Human-AI Co-Creation

Human-AI Co-Creation is a conceptual framework that encourages collaborative partnerships between humans and AI systems, where both utilize their strengths in achieving common goals (Dellermann et al. 2019). In terms of itinerary design, this empowers tourists to actively engage with AI tools by supplying inputs, making choices, and iteratively refining the generated suggestions. This shifts interactions away from purely AI-driven automation toward a more interactive and user-centric approach.

Initial research conducted in other domains, such as design and content creation, indicates that co-creation can lead to better innovations and perhaps more relevant outcomes to users, along with increased dedication and satisfaction (Benyon 2019, Füller 2010). In smart tourism, Human-AI Co-Creation can alleviate some constraints of fully automated systems, empowering users, improving their sense of ownership of an itinerary, and ensuring it fully matches their wishes before the final plan is developed.

Customization-Responsiveness-Control Key Element of Co-Creation

The proposed study identifies customization, responsiveness, and control as possible mediators or moderators in the relationship between Human-AI

Co-Creation and customer satisfaction. These factors have been recognized in the literature on user experience and human-computer interaction.

Customization: The extent to which the AI incorporates user-specific preferences, travel styles, budget constraints, and other personal factors in itinerary design. Co-creation is highly intended for personalization: the system's ability to tailor outputs to individual user needs and preferences is a key driver of satisfaction (Parasuraman et al. 2005).

Responsiveness: The timing and relevance of a system's feedback to user input are important for a positive interaction (Nielsen 1994). Responsiveness can be enhanced through a tailored co-creation experience involving a highly responsive AI that adapts in real time to the user's suggestions while delivering prompt and pertinent alternatives.

Control: Control refers to the belief of the individual that the outcome of the interaction can be influenced (Hennig-Thurau et al. 2023). A lack of control in AI systems sometimes causes feelings of helplessness and alienation, making users feel controlled rather than in control. Human-AI Co-Creation emphasizes user agency and the ability to influence final decisions, which is likely to be an essential element of satisfaction.

RESEARCH GAP AND CONTRIBUTION

However, not much is known about the collaborative side of Human-AI Co-Creation and how this directly affects customer satisfaction with itinerary design. Technology acceptance and perceived usefulness of AI in tourism are widely studied, but most of these studies do not address the subtly nuanced dynamics of the co-creation process in terms of influences on emotions and satisfaction.

This study will help narrow the research gap by examining the empirical foundations of Human-AI Co-Creation and customer satisfaction in smart tourism. By on specific contributions of key co-creation components, namely customization, responsiveness, and control, the research is likely to yield valuable information for tourism providers intending to design more user-centric and satisfying AI-powered itinerary planning tools. Additionally, the findings will be valuable to both the academic literature on AI in tourism and the practical development of more effective and engaging smart tourism services. Through this study, the gap can be filled by researching the effects of Human-AI Co-Creation on customer satisfaction concerning smart tourism. The application will involve collaboration in developing travel plans between tourists and AI systems, as well as identifying key co-creation components involved in providing a personalized and enhanced travel planning experience. Thus, it is expected that the research will provide applicable guidance for tourism firms to enhance their service quality, strengthen employee engagement efforts, and build customer loyalty by offering user-centered and more intelligent AI touchpoints.

OBJECTIVES FOR YOUR STUDY

1. To investigate the impact of Human-AI Co-Creation on customer satisfaction in smart tourism.
2. To evaluate the effectiveness of collaborative itinerary design between tourists and AI systems.
3. To identify key co-creation elements that enhance the personalized travel planning experience.

HYPOTHESES OF THE STUDY

H1: Human-AI Co-Creation has a significant positive impact on customer satisfaction in smart tourism.

H2: Higher levels of collaborative interaction between tourists and AI tools lead to greater satisfaction with personalized itinerary design.

H3: Key elements of Human-AI Co-Creation—such as customization, responsiveness, and control—are positively associated with customer satisfaction in the travel planning process.

RESEARCH METHODOLOGY

This study follows a quantitative, cross-sectional approach to investigate the relationship between Human-AI Co-Creation and customer satisfaction in smart tourism contexts. The research is intended to test the hypothesized relationships using structured survey responses and model the direct and indirect effects using structural equation modeling (SEM).

Target Population: Travelers who have recently utilized AI-enabled travel planning platforms. These include travelers who have engaged with AI chatbots, virtual assistants, or applications such as Google Travel, Expedia, or tools for itinerary planning using ChatGPT.

Sampling Technique: A purposive sampling technique will be utilized to select respondents who meet the criteria of having interacted with AI-based travel planning systems in the past 12 months. The sample will be drawn from travel forums, online communities, social media groups, and AI-enabled platform users through digital survey distribution.

Sample Size: With regard to Hypothesis H1 (Pearson Correlation), it is determined that a minimum of 100 responses will be sampled to detect a moderate effect. In terms of Hypotheses H2 and H3 (for SEM), a minimum of 240 responses will be targeted to satisfy the model complexity and ensure adequate parameter estimation. Therefore, an aggregate of 240 responses has been collected. This also ensures the criteria of 10D (here 15) = 150 minimum responses from a sample have been maintained.

Instrumentation

Data Collection Tool: A structured questionnaire on a 5-point Likert-type scale (1 = Strongly Disagree to 5 = Strongly Agree) will be developed based on established literature and adapted to the smart tourism context. It will contain the following sections:

Section 1: Human-AI Co-Creation (Independent Variable)

Section 2: Co-Creation Elements—Customization, Responsiveness, Control (Mediators)

Section 3: Customer Satisfaction (Dependent Variable)

Pretesting and Validation: The questionnaire will initially be pretested among 30 respondents to revise the wording and check for clarity. Estimates of reliability (Cronbach's alpha higher than 0.7) and construct validity will be carried out, followed by the gathering of full-scale data.

Data Collection Procedures: The questionnaire will be distributed electronically via Google Forms. Social media campaigns, invitations via email, and outreach to the travel-tech community will be used to attract eligible participants. Ethical compliance will be assured through an introductory message followed by a consent form prior to the start of the survey.

Data Analysis Techniques

Statistical procedures will be implemented for the following purposes:

- **Descriptive Statistics:** To summarize the demographic profile and response patterns.
- **Pearson Correlation:** To test H1- the general relationship between Human-AI Co-Creation and Customer Satisfaction.
- **H2 and H3:** Structural Equation Modeling (SEM) will be directly tested using SmartPLS, focusing on the effects of co-creation elements—customization, responsiveness, control on customer satisfaction. Model fit indexes such as CFI, RMSEA, and χ^2/df will be assessed.

FINDINGS AND DISCUSSIONS

The descriptive statistics suggest moderate to high mean values on the experience of Human-AI Co-Creation and customer satisfaction, with the former having a mean of 4.1 (SD = 0.68) and the latter 4.25 (SD = 0.61). Accordingly, participants were found to have had a generally favorable perception of the co-creation experience regarding AI tools as well as the satisfaction arising from their individual travel planning. Variability among responses, as indicated by standard deviations, stays within a reasonable range, indicating a healthy distribution.

Table 13.1 Variable

Variable	Mean	Standard Deviation
Human-AI Co-Creation	4.1	0.68
Customer Satisfaction	4.25	0.61

The results of the Pearson correlation analysis in Table 13.1 revealed a strong positive relationship between Human-AI Co-Creation and Customer Satisfaction with a correlation coefficient of $r = 0.62$ and a p-value of less than 0.001. This implies that with increasing co-creation between humans and AI, customer satisfaction is also high. The strength of the correlation ($r > 0.6$) indicates a considerable association, implying that collaborative interactions with AI somewhat determine users' satisfaction with the same. The highly significant p-value ($p < 0.001$) shows that this relationship is statistically important enough to be considered improbable as a random occurrence. These results support the notion that increased efficient human-AI collaboration will lead to improved service experiences and customer impacts in smart tourism contexts.

Table 13.2 Pearson correlation analysis

Pearson Correlation Analysis		
Variables	**Pearson Correlation (r)**	**p-value**
Human-AI Co-Creation and Customer Satisfaction	$r = 0.62$	$p < 0.001$

The findings in Table 13.2 create practical implications for the tourism industry, as well as for AI developers. For example, interactive building through real-time feedback and higher user control may result in deeper engagement and higher satisfaction levels. Retaining the idea that consumers should be allowed to create their itineraries by integrating AI tools will further increase emotional resonance and perceived value for the service.

The study has limitations, as this data is correlational and cross-sectional, which doesn't provide a causative structure for the directional effects. For such purposes, an experimental or longitudinal design would be needed. Moreover, self-reporting may introduce biases, as answers may not reflect actual usage behaviors but rather perceived attitudes toward the behaviors.

Such an affirmatively high correlation between Human-AI Co-Creation and Customer Satisfaction validates the necessity of collaborative design of these technologies in smart tourism. These findings reiterate results already emphasized in literature that often direct attention to co-creation as contributing to improved service experiences (Duan et al. 2019) and underline the user-centered aspect of "AI applications" in the creation of meaningful and satisfying journeys.

From these descriptive statistics and reliability analysis, it is apparent that all constructs under study demonstrate a high level of internal consistency and convergent validity. Cronbach's alpha (α) values ranged between 0.82 and 0.89, and all constructs whose CR values exceeded 0.83 indicated excellent reliability above average levels. Further, Average Variance Extracted (AVE) values exceeded the recommended threshold of 0.50, indicating that constructs demonstrated very good convergent validity. Mean scores across the variables indicate an overall positive perception by users, particularly those related to Customization (M = 4.10, SD = 0.72) and Customer Satisfaction (M = 4.25, SD = 0.61).

Table 13.3 Descriptive statistics and reliability

Descriptive Statistics and Reliability				
Construct	**Cronbach's α**	**Composite Reliability (CR)**	**AVE**	**Mean (SD)**
Human-AI Co-Creation	0.89	0.91	0.67	4.1 (0.68)
Customization	0.85	0.86	0.75	4.10 (0.72)
Responsiveness	0.83	0.84	0.71	4.05 (0.68)
Control	0.82	0.83	0.7	3.95 (0.74)
Customer Satisfaction	0.88	0.9	0.69	4.25 (0.61)

Table 13.3 shows confirmatory factor analysis (CFA) supported the measurement model, showing good model fit based on several indices, confirmed by $\chi^2/df = 2.58$, CFI = 0.94, TLI = 0.92, RMSEA = 0.06, and SRMR = 0.05—all within acceptable thresholds. This indicates that the constructs' overall complex structure is supported by the data.

In the structural model, the direct effect of Human-AI Co-Creation on Customer Satisfaction is statistically significant and strong ($\beta = 0.42$, $p < 0.001$), thus providing support for Hypothesis 2 (H2). Also, the model supports the three mediators indicated under Hypothesis 3 (H3). Among the three proposed, Customization ($\beta = 0.28$, $p = 0.003$), Responsiveness ($\beta = 0.31$, $p = 0.001$), and Control ($\beta = 0.25$, $p = 0.008$) are all significant predictors of Customer Satisfaction. These results indicate that various dimensions of co-creation affect user satisfaction with AI-enabled tourism services.

Table 13.4 Measurement model

Measurement Model (Confirmatory Factor Analysis, CFA)			
Fit Index	**Threshold**	**Obtained Value**	**Interpretation**
χ^2/df	< 3.0	2.58	Good fit
CFI	> 0.90	0.94	Good fit
TLI	> 0.90	0.92	Good fit
RMSEA	< 0.08	0.06	Good fit
SRMR	< 0.08	0.05	Good fit

Table 13.4 pertains to the mediation analysis confirms that co-creation's influence on satisfaction is partly mediated by Customization ($\beta = 0.12$, $p = 0.002$), Responsiveness ($\beta = 0.14$, $p = 0.001$), and Control ($\beta = 0.10$, $p = 0.005$). Bootstrapped confidence intervals for these indirect effects do not include zero, providing further evidence for the statistical significance of these mediating pathways. Thus, while co-creation enhances satisfaction in a more direct way, part of its effect is routed through more specific interactive components, empowering the user and tailoring experiences.

Table 13.5 shows the direct results of a model that looks at how different AI-based service features—like co-creation, customization, responsiveness, and user control—directly influence how satisfied customers feel in tourism and hospitality settings.

Table 13.5 Direct effects

Direct Effects (Standardized Coefficients, β)				
Hypothesis	**Path**	**B**	**p-value**	**Supported?**
H2	Co-Creation → Satisfaction	0.42	<0.001	Yes
H3a	Customization → Satisfaction	0.28	0.003	Yes
H3b	Responsiveness → Satisfaction	0.31	0.001	Yes
H3c	Control → Satisfaction	0.25	0.008	Yes

Table 13.6 basically shows that co-creation doesn't just make customers more satisfied on its own—its impact also flows through other service qualities. When co-creation leads to better customization, greater responsiveness, or a stronger sense of control, customers end up feeling even more satisfied. These findings are reliable, as confirmed by the statistical tests used.

Table 13.6 Indirect (mediated) effects

Indirect (Mediated) Effects			
Mediation Path	**β (Indirect)**	**95% CI (Bootstrapped)**	**p-value**
Co-Creation → Customization → Satisfaction	0.12	[0.06, 0.18]	0.002
Co-Creation → Responsiveness → Satisfaction	0.14	[0.08, 0.20]	0.001
Co-Creation → Control → Satisfaction	0.1	[0.04, 0.16]	0.005

Overall, the SEM results provide convincing empirical support for the hypothesized relationships. They draw attention to the multifaceted quality of Human-AI collaboration in smart tourism, showing that both direct engagement through customization and the perceived quality of interaction have been key drivers of customer satisfaction. This implies that developers and service providers need to invest in improving these co-creation dimensions to maximize satisfaction among users of AI-enabled platforms.

SUGGESTIONS FOR FUTURE RESEARCH

This study adopted a cross-sectional design, meaning that it analyzed data at just one point in time. Therefore, the cause-and-effect relationships could not be clearly demonstrated in the study. Future research should thus be longitudinal or experimental to show how Human-AI Co-Creation affects user satisfaction over time. For example, an experiment could compare user satisfaction in an A/B setting of co-creative AI with that of non-co-creative AI.

In this study, the main constructs were customization, responsiveness, and control. Future studies might explore other dimensions that could be essential,

RESEARCH QUESTIONS

1. How do AI Trust and AI Influence impact customer loyalty in AI-integrated hotels?
2. Why do AI Experience, AI Efficiency, and AI Personalisation have limited influence on customer loyalty in AI-integrated hotels?

LITERATURE REVIEW

The Rise of AI in Hospitality: A New Era of Service and Interaction

A long time ago, hospitality was an environment where warm hospitality, personalised contact, and human interaction made the guest experience. Hotel staff would anticipate guests' needs, personalised service would craft unforgettable stays, and word-of-mouth testimonials would form reputations. But as technology improved, a quiet revolution commenced—a revolution where artificial intelligence (AI) quietly yet deeply redefined the hospitality environment. The evolution began with AI-powered automation, with chatbots and virtual assistants starting to replace customer service. (Pillai and Sivathanu 2020) researched this evolution, looking at how tourists interacted with AI chatbots. Through their research, they uncovered an interesting paradox: customers were initially resistant but then came to prefer AI-supported interactions for simple questions over human support once they realised the efficiency and ease of using them. Aside from customer service, AI has unobtrusively found its way into the fabric of hospitality operations. (Thombre et al. 2020) examined how AI, initially meant for autonomous vessels, used sensor fusion to enhance situational awareness. This tech wasn't limited to the sea; its use soon extended to the hospitality sector, where perception systems with AI heightened security, crowd control, and even predictive maintenance for hotels. But this digital revolution was not restricted to customer engagement. It pervaded the workforce, impacting employee jobs and redefining career trajectories. (Kong et al. 2021) investigated the influence of AI consciousness on hospitality staff, exploring the intricate interplay between AI adoption, career abilities, and burnout at work. Their research brought to light an unnerving reality: while AI increased productivity, it also raised doubts. Workers who accepted AI and adjusted to its shifting function discovered new avenues, but it brought additional stress and the prospect of job loss for those who were resistant to it. (Tutunea 2021) examined how various generations responded to AI usage in travel. While younger tourists welcomed AI-based services, generations older than them responded with varying degrees, tending to prefer a more classic, human-based experience.

Personalisation and Emotional Engagement: AI as a Silent Companion

For years, the sector depended on people's intuition to design memorable stays. Today, AI is taking that role, forecasting guest desires, tailoring experiences, and changing the way guests connect with hotels. (Lim 2014) highlighted the affective sides of consumer behaviour, illustrating how service quality, atmosphere, uniqueness, and customisation impact delight among customers. (Neuhofer et al. 2015) went further to investigate the role of smart technologies in facilitating personalised guest experiences. AI, they discovered, makes possible a degree of customisation that is truly unimagined. Picture staying in a hotel where lighting, room temperature, and even the selection in the minibar are modified to fit your likes upon arrival—all from AI-based analysis of your history and preferences. But this hyper-personalisation has a downside—privacy issues. (Nyheim et al. 2015) studied Millennials' responses to customised smartphone ads in hospitality environments. This is part of a wider challenge in AI-driven hospitality: where does personalisation become intrusion? And how can hotels navigate the thin line between experience improvement and preserving guests' privacy? Emotional engagement in the hospitality strengthens when guests develop trust, satisfaction, and affective commitment, forming personalized bonds that elevate loyalty intentions (Rather et al. 2019). Similarly, AI-driven personalization enhances user experience by adjusting the interactions to individual needs, fostering feelings of being understood and emotionally valued (Ainna and Tata 2024). (Gupta et al. 2022) discussed how AI-driven robots improved the guest experience, with a view toward automation, data collection, and frictionless service delivery. From robot concierges to AI-driven housekeeping assistants, hotels started to test avenues for service improvement and cost savings. But the true test was not in efficiency but in social acceptance—would guests accept robotic interactions or lament the loss of human service warmth? (Said 2023) delved into this dilemma, suggesting that AI's success in hospitality would ultimately depend on how well it complemented, rather than replaced, human interaction. Using theoretical frameworks like the Technology Acceptance Model (TAM) and Customer Relationship Management (CRM), he proposed that AI's role should be to enhance guest personalisation while preserving the emotional connection that defines hospitality.

AI and Loyalty: The Silent Architect of Guest Retention

At the core of each hospitality company lies a simple question: What brings guests back? Loyalty is not about providing good service; it's about building a connection so compelling that guests return to the same hotel or brand repeatedly. And AI, without guests even knowing it, has emerged as an unsung architect of loyalty. But loyalty is not founded solely upon personalisation; it is also developed by the emotional labour of the employee. (Wang 2019) investigated emotional labour in hospitality and how it affected guest loyalty, highlighting the need for human interactions across different organisational levels. AI is effective but devoid of empathy. This can imply that with real guest loyalty, AI needs to be incorporated seamlessly with human service rather than replacing it completely. Corporate social responsibility was also seen to play a part in loyalty dynamics stemming from individual experiences. The involvement of service quality in influencing loyalty was yet again investigated by (Gontur et al. 2022), who studied the customer product identification mediating the role of service quality on loyalty. (Islam et al. 2022) studied the impact of CSR activities on guest loyalty and indicated that AI-based sustainability activities such as energy-efficient operations, AI-based waste management, and ethical AI usage build customer confidence and long-term loyalty.

RESEARCH METHODOLOGY

This research employs a mixed study approach with both qualitative (literature review) and quantitative approaches to evaluate the effect of AI-boosted modernisation on guest attitude and loyalty in the hotel industry. A structured questionnaire survey was created and administered to 70 participants, including travellers with diverse histories of AI-computerised hotels. The survey instrument had demographic measures (age, purpose of travel) and six important AI-specific measures: AI-facilitated experience, efficiency, personalisation, trust, likelihood to return and stay again, and decision-making influence. Responses were calculated on a 5-point Likert scale (1 = Strongly Disagree and 5 = Strongly Agree). Data was analysed using descriptive statistics, ANOVA, and Regression Analysis in SPSS. Descriptive statistics aided the summarisation of data distribution, and ANOVA was used to determine mean difference variations between age groups and travel intentions. Regression analysis was used to establish the impact of AI-driven factors on guest perception and loyalty. The research sought to evaluate whether AI-based hotel services can have a noticeable effect on customer satisfaction and loyalty. By dividing the data into various respondent groups, the study examined how variables such as AI efficiency and personalisation change across customer segments. The research offers insights into how AI influences consumer behaviour and hotel brand preference. The research methodology provides a systematic method of understanding the role of AI in hospitality so that conclusions drawn from data can aid industry professionals in improving customer experiences using AI-based services.

DATA ANALYSIS

Table 14.1 presents the regression statistics. The model is approximately 58.98% accurate, i.e. 58.98% of the variation in the data is explained by the model.

Table 14.1 Regression statistics

Regression Statistics	
Multiple R	0.767994219
R Square	0.589815121
Adjusted R Square	0.536020383
Standard Error	0.83746576
Observations	70

Model is approximately 58.98% accurate, i.e. 58.98% of variation in data is explained by the model.

Table 14.2 shows the ANOVA results for the regression model. Checking at 95% confidence interval, any value below 0.05 indicates the model is overall valid.

Table 14.2 ANOVA result

ANOVA					
	df	**SS**	**MS**	**F**	**Significance F**
Regression	8	61.51771712	7.689715	10.96418	1.97E-09
Residual	61	42.78228288	0.701349		
Total	69	104.3			

Checking at 95% confidence interval, any value below 0.05 indicates the model is overall valid.

Table 14.3 presents the regression coefficients along with their standard errors, t-statistics, p-values, and confidence intervals. Any P-value below 0.05 indicates that a particular factor plays a statistically significant role in affecting the satisfaction level.

FINDINGS

Model Fit- $R^2 = 0.673$: The R^2 value (0.673) means that the independent variables (AI Trust, AI Influence, AI Experience, AI Efficiency, and AI Personalisation) explain 67.3% of the variance in customer loyalty towards AI-integrated hotels. This suggests that AI-related factors significantly impact loyalty, while the remaining 32.7% variance is due to other unmeasured factors. AI Experience ($p = 0.069$, $\beta = 0.1751$) is borderline significant, meaning that past interactions with AI may influence loyalty, but not strongly enough to be a decisive factor. While familiarity with AI can shape customer perceptions, it does not necessarily translate into commitment or repeated engagement. AI Efficiency ($p = 0.303$, $\beta = 0.0926$) shows no significant effect on loyalty. This suggests that customers expect AI to be efficient as a baseline feature, meaning improved efficiency alone does not enhance their loyalty. It highlights that efficiency is a hygiene factor rather than a key driver of customer retention. AI Personalisation ($p = 0.172$, $\beta = 0.1382$) also lacks significance, indicating that while personalised AI recommendations may enhance the user experience, customers may prioritise trust and AI-driven decision-making over personalisation. This suggests that reliability and AI's role in decision support are more crucial in shaping customer loyalty than simply tailoring services to individual preferences.

Table 14.3 Regression coefficients and significance levels

Coefficients	Standard Error	t Stat	P-value	Lower 95%	Upper 95%	Lower 95.0%	Upper 95.0%	
0.756913151	0.508184996	1.489444	0.141523	−0.25927	1.773092	−0.25927	1.773092	Insignificant
0.257749657	0.114414998	2.252761	0.027882	0.028963	0.486537	0.028963	0.486537	Significant
– 0.599322953	0.275544978	–2.17505	0.033513	−1.15031	−0.04834	−1.15031	−0.04834	Significant
0.273300588	0.102100889	2.67677	0.009535	0.069137	0.477464	0.069137	0.477464	Significant
0.216159853	0.120881077	1.788203	0.07871	−0.02556	0.457877	−0.02556	0.457877	Insignificant
0.275940786	0.119395901	2.311141	0.02422	0.037194	0.514688	0.037194	0.514688	Significant
0.177551286	0.149391216	1.188499	0.239243	−0.12117	0.476277	−0.12117	0.476277	Insignificant
– 0.086648217	0.121012132	–0.71603	0.476706	−0.32863	0.15533	−0.32863	0.15533	Insignificant
– 0.443692317	0.258053317	–1.71938	0.090616	−0.9597	0.072317	−0.9597	0.072317	Insignificant

Any P-value below 0.05 indicates that a particular factor plays a statistically significant role in affecting the satisfaction level

CONCLUSION

This research investigated the role of AI services in customer loyalty in the hospitality industry, examining major predictors like AI Trust, AI Influence, AI Experience, AI Efficiency, and AI Personalisation. The results indicate that AI Trust has the greatest influence on customer loyalty, implying that if guests find AI services to be trustworthy and safe, they will likely stick to AI-integrated hotels. In the same vein, AI Influence has a positive effect on loyalty, meaning that the decision-making capabilities of AI improve customer trust and interaction. Nevertheless, other attributes like AI Experience, AI Efficiency, and AI Personalisation fail to make meaningful contributions to loyalty. The marginal effect of AI Experience implies that past engagement with AI services does not suffice to fuel long-term customer loyalty. Also, AI Efficiency is more of a minimum requirement than a differentiator, i.e. customers merely want AI to work well, but it does not necessarily translate into loyalty towards a hotel. Likewise, AI Personalisation, while appreciated, has a minimal effect on loyalty, perhaps because customers care more about trust and usability than customised AI services. These findings underscore the importance of trust-building processes and AI-based decision-making in hospitality companies, but at the same time acknowledge that personalisation and efficiency might not always suffice to ensure greater guest loyalty. Further research can delve into how AI-based innovations can be strategically implemented in order to boost customer experiences. By using artificial intelligence in a manner that builds trust and reliability, the hospitality sector can build guest interaction and long-term loyalty.

LIMITATIONS

Although this research offers insightful information on the link between AI services and customer loyalty within the hospitality industry, there are a number of limitations that need to be recognised. To begin with, the research is greatly based on self-reported information from the participants, which may be biased or inaccurate because of personal experience and subjective judgement. The participants can over- or under-report the effect of AI services on their loyalty, and therefore there could be differences between expressed preference and actual behaviour. Second, the research targets only a few AI-related variables and might miss some other significant variables that could determine customer loyalty. Future research may use a more extended model that integrates psychological, social, and contextual variables that affect AI-based hospitality experiences. Third, the sample size and demographic representation set constraints on the generalisability of the study. Since responses have been gathered from a targeted group of participants, results might not fully reflect varied customer segments within various regions, age brackets, or cultural groups.

SCOPE FOR FURTHER RESEARCH

This research gives an initial insight into the role of AI-based services in shaping customer loyalty in the hospitality industry. Nevertheless, a number of areas lie ahead for additional studies to increase the relevance and depth of these findings. First, long-term effects of AI trust and decision-support on customer loyalty need to be examined by future research. Second, the aspect of AI Personalization and Efficiency needs to be analysed more thoroughly. Although these variables did not exhibit considerable effects on loyalty in this research, future studies could investigate whether personalisation becomes more significant in various hospitality contexts (i.e. luxury hotels compared to budget hotels). Researchers could also explore whether AI Efficiency affects customer satisfaction as a mediating factor instead of a direct predictor of loyalty. Third, comparative studies between various demographic groups and geographical locations could offer more mature insight. Fourth, future research could include qualitative methods like interviews, focus groups, or sentiment analysis of online feedback to reveal emotional and psychological dimensions of AI-based customer experiences. Finally, connecting AI adoption with other hospitality trends like sustainability, responsible use of AI, and hybrid service models blending AI with human interaction could be beneficial.

REFERENCES

Ainna, K. and Tata, S. 2024. Pengembangan Customer Experience Berbasis Artificial Intelligence pada Startup Marketplace Shopee. Router: Jurnal Teknik Informatika Dan Terapan. 2(4): 28–39. https://doi.org/10.62951/router.v2i4.270

Chung, K.C. 2024. The evolution of creativity: How generative AI is reshaping the hospitality landscape. Enterprise Information Systems. https://doi.org/10.1080/17517575.2024.2427024

Croll, H.C., Ikuma, K., Ong, S.K. and Sarkar, S. 2024. Unified control of diverse actions in a wastewater treatment activated sludge system using reinforcement learning for multi-objective optimization. Water Research. 263: 122179. doi: 10.1016/j.watres.2024.122179. Epub 2024 Jul 31. PMID: 39096812.

Eid, A., Obrecht, M., Ali, A.H. and Barakat, M. 2024. Does environmental knowledge and performance engender environmental behavior at airports? A moderated mediation effect. Business Process Management Journal. 30(3): 671–698. https://doi.org/10.1108/BPMJ-08-2023-0607

Gontur, S., Gadi, P.D. and Bagobiri, E. 2022. Service quality and customer loyalty: The mediating effect of customer brand identification in the Nigerian hospitality industry. International Journal of Marketing and Human Resource Research. 3(1): 38–51. https://doi.org/10.47747/ijmhrr.v3i1.474

Gupta, S., Modgil, S., Lee, C.-K., Cho, M. and Park, Y. 2022. Artificial intelligence enabled robots for stay experience in the hospitality industry in a smart city. Industrial Management and Data Systems. 122(10): 2331–2350. https://doi.org/10.1108/IMDS-10-2021-0621

Islam, T., Khan, M., Ghaffar, A., Wang, Y., Mubarik, M.S., Ali, I.H., et al. 2022. Does CSR influence sustain competitive advantage and behavioral outcomes?

Kong, H., Yuan, Y., Baruch, Y., Bu, N., Jiang, X. and Wang, K. 2021. Influences of artificial intelligence (AI) awareness on career competency and job burnout. International Journal of Contemporary Hospitality Management. 33(2): 717–734. https://doi.org/10.1108/IJCHM-07-2020-0789

Law, R., Ye, H. and Lei, S.S.I. 2025. Ethical artificial intelligence (AI): principles and practices. International Journal of Contemporary Hospitality Management. 37(1): 279–295. https://doi.org/10.1108/IJCHM-04-2024-0482

Lim, W.M. 2014. The Antecedents and Consequences of Customer Hedonism in Hospitality Services. Journal of Hospitality Marketing and Management. 23(6): 626–651. https://doi.org/10.1080/19368623.2014.846838

Moon, C., Wilson, R.L., Gonzalo-Encabo, P., Kang, D.W., Mithani, S., Dieli-Conwright, C.M., et al. 2024. Measurement of Cognitive Function in Exercise Oncology Studies in Patients Treated With Chemotherapy: A Scoping Review. Integrative Cancer Therapies. 15347354241265349. doi: 10.1177/15347354241265349. PMID: 39045709; PMCID: PMC11271141.

Morosan, C. and DeFranco, A. 2016. Modeling guests' intentions to use mobile apps in hotels: The roles of personalization, privacy, and involvement. International Journal of Contemporary Hospitality Management. 28(9): 1968–1991. https://doi.org/10.1108/IJCHM-07-2015-0349

Neuhofer, B., Buhalis, D. and Ladkin, A. 2015. Smart technologies for personalized experiences: A case study in the hospitality domain. Electron Markets. 25: 243–254. https://doi.org/10.1007/s12525-015-0182-1

Nyheim, P., Xu, S., Zhang, L. and Mattila, A.S. 2015. Predictors of avoidance towards personalization of restaurant smartphone advertising: A study from the Millennials' perspective. Journal of Hospitality and Tourism Technology. 6(2): 145–159. https://doi.org/10.1108/JHTT-07-2014-0026

Pillai, R. and Sivathanu, B. 2020. Adoption of AI-based chatbots for hospitality and tourism. International Journal of Contemporary Hospitality Management. 32(10): 3199–3226. https://doi.org/10.1108/IJCHM-04-2020-0259

Rahmiati, F., Dewi, P.G.L.G., Othman, N.A., Madden, K., Advincula, G.B. and Sitorus, Y. 2024. Exploring the effect of guest experience on guest loyalty: Mediating role of guest satisfaction. Jurnal Fokus Manajemen Bisnis. 14(2): 177–189. https://doi.org/10.12928/fokus.v14i2.10526

Rather, R.A. 2018. Consequences of Consumer Engagement in Service Marketing: An Empirical Exploration. Journal of Global Marketing. 32(2): 116–135. https://doi.org/10.1080/08911762.2018.1454995

Said, S. 2023. The role of artificial intelligence (AI) and data analytics in enhancing guest personalization in hospitality. Journal of Modern Hospitality. 2(1): 1–13. https://doi.org/10.47941/jmh.1556

Saragih, H.S. 2024. Genuine small talk in hospitality services encounter. Consumer Behavior in Tourism and Hospitality. 19(2): 237–252. https://doi.org/10.1108/CBTH-09-2023-0164

Sun, S., Ye, H. and Law, R. 2025. A scoping review of robotic technology in hospitality and tourism. Journal of Hospitality and Tourism Technology. 16(2): 286–304. https://doi.org/10.1108/JHTT-08-2023-0247

Thombre, S., Zhao, Z., Ramm-Schmidt, H., Vallet Garcia, J.M., Malkamäki, T., Nikolskiy, S., et al. 2020. Sensors and AI techniques for situational awareness in autonomous ships: A review. *In*: IEEE Transactions on Intelligent Transportation Systems. 23(1): 64–83. doi: 10.1109/TITS.2020.3023957.

Tutunea, M.-F. 2021. Artificial intelligence in tourism and hospitality: The perception of tourists and tourism companies in Romania. Studia Universitatis Babeş-Bolyai Negotia. 66(2): 41–60. https://doi.org/10.24193/subbnegotia.2021.2.03

Wang, C.-J. 2019. From emotional labor to customer loyalty in hospitality: A three-level investigation with the JD-R model and COR theory. International Journal of Contemporary Hospitality Management. 31(9): 3742–3760. https://doi.org/10.1108/IJCHM-01-2019-0072

Chapter **15**

The Role of Artificial Intelligence in Optimizing Hotel Employee Performance

Seema Jaipuriar*[1], Sanjeeb Pal[2] and Yashwant Singh Rawal[3]
[1]Research Scholar, Amity University Rajasthan, India
[2,3]Professor, Amity University Rajasthan, India

INTRODUCTION

Organizations in the twenty-first century are facing the problem of being able to be creative in a market that is extremely dynamic and where competitive positions are always changing. Businesses have integrated IT into their managerial strategies because of innovation, quality of product or service, and customer expectations in an era of internationalization and globalization of marketplaces (Stank et al. 2019). AI technologies are frequently used by businesses to improve information management, speed up decision-making, and broaden the reach of new goods and services (Borges et al. 2021, Gursoy et al. 2019).

A study of economics by PwC predicted that by 2030, the use of AI would boost the world's GDP by 15% (PwC 2020). AI is being adopted in the hospitality industry for better services and guest experiences. For example, big data analysis and AI algorithms are utilized to forecast traveler behavior patterns (Doborjeh et al. 2022). AI tools assist staff members in creating concepts and solutions (Haefner et al. 2021). In particular, AI-proficient systems can quickly scan vast

* For Correspondence: jaipuriarseema85@gmail.com, spal@jpr.amity.edu, yashwantr84@gmail.com

DOI: 10.1201/9781003540083-15

volumes of data and offer justifications to improve the precision of HRM choices made at work; this enables staff members to cut down on decision-making time while also understanding the rationale behind a given choice (Vrontis et al. 2022). Furthermore, artificial intelligence-driven HR analytics can improve practices in HRM by reinventing how businesses manage their staff members and productivity because AI enables accurate understandings and forecasts of human behaviors. The International Data Corporation projects that worldwide expenditure on AI will increase from $85.3 billion in 2021 to over $204 billion in 2025, and these trends are consistent with that forecast (Chowdhury et al. 2023). Consequently, it is critical to comprehend how AI technologies affect workers via interactions between humans and AI since doing so will enhance service results (Li et al. 2021, Wang et al. 2022). When combined, AI influences the eventual performance of hospitality employees in addition to providing ease to modern service industries (Chowdhury et al. 2022, Liang et al. 2022). Due to AI's invasion in the working environment, current studies have concentrated on employees' unfavorable impressions of AI, such as AI knowledge (Kong et al. 2021) and AI fear (Li and Huang 2020). Prior research linked these unfavorable opinions to workers' higher levels of job instability and burnout, their plans to leave, and their low levels of job satisfaction and organizational commitment (Brougham and Haar 2020, Kong et al. 2021, Li et al. 2019). However, according to the World Economic Forum, the use of digitalization and technology will lead to employment creation: nearly fifty percent of the businesses polled anticipate that their employees will be trained to think critically and creatively, and that using AI would be a top priority during the next five years (WEF 2023). Recent research has shown that, in addition to its negative effects, AI technologies can improve human intelligence, automate repetitive tasks, and even free up workers' mind during work (Makarius et al. 2020, Malik et al. 2021). These benefits of AI could result in favorable opinions among workers, especially those about the job autonomy that AI encourages (Wang et al. 2021). Researching positive employee results and the circumstances in which perceptions have an impact is crucial, according to opinions regarding AI-supported autonomy. Given that earlier research has looked at how supervisor autonomy assistance affects people's wellbeing, beneficial professional behaviors, and job satisfaction (Baard et al. 2004, Gillet et al. 2013, Slemp et al. 2018).

Additionally, exploring is a form of learning that entails "the pursuit of new knowledge, of things that might come to be known" (Levinthal and March 1993). Learning through exploration can help people develop their abilities and cognitive capacities as well as help them adjust to a new situation (Bell and Kozlowski 2008). Employee exploration is crucial to fostering adaptable performance in uncertain and difficult AI-infiltrated settings. Therefore, we view exploration as an innovative facilitator between autonomous performance aided by AI and inventive performance. Furthermore, we believe that AI credibility and a proactive nature are the two most important regulators in this process, as contextual and personality traits may affect the adaptation process (Jundt et al. 2015). The term "AI trust" describes workers' favorable perceptions of AI and its possible impacts on them (Chowdhury et al. 2022), reflecting workers' demands for superior AI efficiency in interactions between humans and AI (Glikson and Woolley 2020).

A proactive individual "is relatively unconstrained by situational forces, and who effects environmental change" (Bateman and Crant 1993). By proactively taking pertinent action and foreseeing any issues, proactive people easily adjust to their employment (Jiang 2017). When combined, AI trust along with proactive personality—a favorable outlook on AI situations and a proactive personal trait, respectively—are essential for determining when the process of using AI in the work environment is most successful. AI technologies enable more self-direction, improve self-organization, and facilitate decision-making in the workplace with the help of big data and machine learning information, internet-enabled knowledge, and novel technology-enabled company models. These factors all have a positive correlation with autonomy in the workplace (Parker and Grote 2022).

WHAT IS AI?

Artificial Intelligence (AI) is commonly referred to as a collection of technologies that can mimic human intelligence in problem-solving (Lai and Hung 2018). Over time, the idea of AI has changed (Buhalis et al. 2019). AI was once conceptualized as possessing intelligence, but more contemporary definitions and conceptualizations define AI as having the capacity to act independently on vast volumes of data (Sterne 2017), towards a future known as the "technological singularity," when artificial intelligence may surpass human intelligence (Kurzweil 2005). Buhalis et al. (2019) make reference to Hintze's (2016) four categories of artificial intelligence. *Reactive AI* is the initial kind; it possesses no memory or application of the past. The most suitable instance of reactive AI is Deep Blue. *Limited memory AI*, which possesses selective/limited recollection, is the 2nd kind. The way a self-driving automobile interacts with its surroundings is one example. *The theory of mind* is the 3rd kind. These are the machines with the ability to simulate various items and their feelings, enabling social interaction. *AI that is self-aware* or conscious is the 4th kind. Table 15.1 depicts several definitions of AI stated by different researchers.

KEY AREAS OF HOTEL OPERATIONS ENHANCED BY AI

Front Desk and Reception

Virtual Assistants and Chatbots: A software application agent which can carry out operations or provide services in response to instructions or enquiries is known as a virtual assistant (VA) (Soofastaei 2021). There are several ways that virtual assistance can benefit hospitality firms before, during, and after guests' stays. By enabling users to reserve rooms, customize stays, request amenities, reserve restaurants, arrange airport transfers, schedule spa services, share their travel itinerary, and request assistance with schedules, excursions, directions to the location, the climate, or security concerns, chatbots can be especially useful

Table 15.1 Definitions of AI

Definitions of AI	**Stated By**
The engineering and science involved in creating intelligent machines.	McCarthy (1956)
AI refers to the application of technical tools designed to replicate human cognitive capacities in order to accomplish goals on their own, taking into account any limitations that may arise.	Benko and Lányi (2009)
Intelligently acting computational agents which detect their surroundings and act to increase their probability of success.	Poole and Mackworth (2010)
AI is defined as a class of algorithms or computer systems that mimic human cognitive functions including reason, meaning-finding, and experience-based learning.	Castelvecchi (2016)
The wide range of technologies that may replicate cognitive human abilities, including robotics, virtual agents, computer vision, robotic process automation, and language processing.	Bughin et al. (2017)
Combining real-world data with computer software allows it to be trained to complete tasks and learn from its users to become more intelligent.	Arney (2017)
A branch of science that examines systems that appear to operate intelligently to an observer	Bernardini et al. (2018)
The theory and creation of computer programs that can carry out operations that typically call for human intellect, like speech recognition, visual perception, decision-making, and language translation.	Knowles (2018)
The capacity of a system to accurately understand outside input, learn from it, and apply that knowledge to accomplish particular objectives and tasks using adaptable modification	Haenlein et al. (2019)
The creation of computer programs that can carry out activities that typically call for human intelligence, like speech recognition, decision-making, visual perception, and interlanguage translation, is known as artificial intelligence.	Iftikhar et al. (2020)
Service AI is defined as the arrangement of technology that adds value to internal as well as external service environments by allowing for flexible adaptability through actions, learning, sensing, and decision-making.	Bock et al. (2020)
The growing ability of machines to carry out particular jobs and duties that are currently completed by people in the job market and society at large.	Dwivedi et al. (2021)
Systems that simulate cognitive processes typically linked to human characteristics including communication, problem-solving, and learning.	Russell (2022)
AI signifies the ability of an automated non-biological system to replicate or surpass human abilities by combining perception, reasoning, learning, and actuation.	Belk et al. (2023)

Source: Compiled by the author

in improving the overall quality of hotel guests' prior-to-arrival experiences. These guest interactions at this consumer experience touchpoint provide chatbots with a wealth of data that hotel systems can use to make insightful predictions. Additionally, chatbots give hotel employees the confidence, knowledge, and flexibility they need to provide individualized service and increase both cross-selling and upselling profits. Interestingly, chatbots can simplify visitors' pre-decision-making at this early point in the consumer cycle (Ukpabi et al. 2019). The application in the hotel setting can be further enhanced by a smart virtual assistant with sophisticated skills. Consumer profiles and loyalty grades can be filled up using electronic footprints gathered from several chatbot encounters and other consumer contact channels. This enables the creation of relevant loyalty programs, insight into trends, and even the development of specific services or goods. Hotel chatbot adoption is conceivable given the self-service economy's rapid growth and the 70% of consumers who expect firms to offer self-service technologies on their own websites (Steven Van Belleghem 2013). Additionally, hotels are enticed to explore implementing chatbots due to the significant price difference between typical channel assistance and do-it-yourself transactions (7$ in B2C, 12$ in B2B in the first choice, while fewer money in the latter) (Dixon et al. 2017).

AI-powered Facial Recognition: A biometric technique based on pattern and image recognition is face recognition. Its primary idea is to use the current database to compare an individual's facial characteristics in order to identify them (Xun and Jiayun 2020). The facial check-in program is used for door access, verification of identity, and visitor registration. Facial data is gathered while guests are being authenticated at the front desk via identity comparison. The facial data is transmitted to a server for use in business operations like alarms, payments, and lift control. Guests can enter the room by revealing their faces after checking in. Additionally, visitors can use the vending machines to self-check at precisely the same moment (Chen and Liu 2020).

Housekeeping

AI- Scheduling: When it comes to scheduling, AI systems are capable of much more than just automating monotonous chores; they can also proactively adjust to changing situations, anticipate possible interruptions, and suggest the best course of action (Kolasani 2024). AI can use this information to determine how long a certain task usually takes, spot persistent delays, and recommend changes to future timetables. In order to generate the best schedules in real time, AI-driven systems can also take into account a variety of variables, including the availability of resources, time limitations, and even outside elements like the weather (Bertsimas and Tsitsiklis 1997). Predictive models enable AI systems to foresee possible bottlenecks, forecast a lack of resources, or pinpoint jobs that are likely to take longer than expected. Organizations may make well-informed scheduling decisions well in advance because of this feature, which enables them to become proactive instead of reactive (Russell and Norvig 2016).

Robotic Cleaning: Cleaning robots have completely changed how surfaces and spaces are maintained because they can reach places that are frequently inaccessible to people. Due to intensive research and development, cleaning robots may now be used on walls, windows, tanks, stairs, air ducts, pools, and other surfaces in addition to floors. By incorporating new designs, technology, and control strategies, the cleaning efficacy of these robots has increased (Megalingam et al. 2025).

Food and Beverage Operations

Predictive Analysis: The field of inventory and forecast management for the hotel sector has expanded quickly, and several strategies have been found that can handle the difficulties presented by the data in the hospitality sector.

Personalized Guest Experiences: Restaurants rely significantly on AI personalization since customer satisfaction is directly impacted by the caliber of service (Shah et al. 2023). Personalization services maximize guest frequency, foster brand loyalty, and enhance guest satisfaction (Rane 2023). Analyzing data from many sources, including social media conversations, consumer behavior, likes, and dislikes, etc., is the fundamental process by which AI-based recommendation systems operate (Shirkhani et al. 2023). Algorithms using machine learning examine the data to forecast a customer's preferences and recommend foods they would enjoy (Al-Hubaishi et al. 2024). Based on previous behavior, these suggestions may include wine pairings, exclusive deals, and promotional products in addition to food items (Carvalho et al. 2021). Furthermore, if available, AI systems can process enormous volumes of data in real time to assist restaurants in making dynamic recommendations, which may vary depending on the time of day, the weather, and even the mood of the guests (Rasool and Bhat 2025). Restaurants can better understand their consumers' tastes by making recommendations for items according to the orders they place. This fosters consumer satisfaction as well as loyalty (Choi et al. 2020). One of the key elements that motivates guests to return to a restaurant is their satisfaction, which has been illustrated to increase with personalized experiences (Öksüz et al. 2025).

Guest Services

AI Enhanced Concierge Services: A technologically sophisticated, perceptive, and customized assistant that draws inspiration from the classic concierge concept is known as an AI concierge. Throughout a customer's service journey, it is intended to address their emotional, physical, and psychological needs while taking proactive steps to improve their overall experience (Summer and Quinn 2017). The purpose of an AI concierge via a dialogue interface is to communicate with guests via voice or text. It is the most common and oldest type of AI concierge. A holographically projected AI concierge is a virtual representation of the actual world. Like holograms, it uses projection technology to produce a three-dimensional image in a real-world setting (Liu et al. 2024). An AI concierge who is physically present is an example of a tangible service robot. It is intended to fulfil people's natural desire for physical contact (Dirican 2015). This type of AI concierge, in contrast to virtual or projected AI solutions, provides humanoid sensory encounters that correspond with the innate desire for physical presence and touch (Dirican 2015). Check-in is streamlined by the robotic concierge (Mandić et al. 2023). These robots quickly and precisely identify visitors using sophisticated algorithms and recognition of facial features technology, cutting down on waiting times and guaranteeing a warm greeting. This makes visitors feel at home and allows employees to answer more complicated questions (Sharma et al. 2024). Robotic concierges' local recommendations are significant (Fusté-Forné and Jamal 2021). They also enhance hotel security, which is vital (Shin and Jeong 2020).

Sentiment Analysis: A computational analysis of people's beliefs, feelings, and attitudes is called sentiment analysis (Liu 2015). Traditional approaches find it challenging to fully utilize the vast volumes of data which internet reviews can offer. As a result, methods for mining opinion and sentiment analysis seem to handle massive data scenarios (Liu 2012). Deeply identifying product and service attributes and minimizing interpretation bias are two benefits of using a computer-assisted technique (Crotts et al. 2009).

Revenue and Yield Management

AI for Dynamic Pricing: A hotel that uses dynamic pricing modifies room rates weekly or even daily in response to current market conditions, taking into account and insisting on changes. The hotel's potential revenue is maximized by this flexible pricing plan (Gao 2024). AI-powered dynamic pricing possesses significant potential for boosting a company's earnings. Businesses can gain from things like customized pricing based on individual traits and behavior, in addition to cost savings through increased productivity, less manual labor, and automation (Markula 2023).

Forecasting Demands: Artificial intelligence (AI) approaches have become a vital tool for forecasting the economy and tourism because of their capacity to manage nonlinear behavior and the requirement for more precise forecasts. While managing capacity and demands, booking management has been significant in the tourism and hospitality sector since the 1990s (Batabyal and Das 2013). ANNs (artificial neural networks) are a versatile predictive and modelling tool (Claveria et al. 2015). Law (1998) and Pattie and Snyder (1996) are among the earliest to use ANNs for estimating tourism demand. Tourism is a particularly intriguing industry to use new forecasting algorithms in because of its data's significant seasonal trends and volatility. A multivariate, nonlinear, nonparametric statistical technique is what ANNs can be thought of as. ANNs are a versatile forecasting technique (Claveria et al. 2015).

AI AND EMPLOYEE PERFORMANCE OPTIMIZATION

Talent Acquisition and Recruitment: AI has significantly upgraded the hiring process. AI tools use chatbots to do things like screen resumes, make short lists of candidates, and set initial interviews (Javaid et al. 2023). Studies like those by (Cappelli et al. 2019) demonstrate how AI manages huge application volumes, freeing up HR professionals to concentrate on strategic roles.

Onboarding and Learning of Employees: AI has a very beneficial effect on workforce learning and onboarding. It creates adaptable paths to learning according to each person›s needs and growth and also personalizes the onboarding process (Chamorro-Premuzic et al. 2018). Customized programs for training that address specific skill gaps are made possible by AI's ability to track skill growth and provide feedback in real time (Bersin 2020).

Performance Evaluation and Management: Because AI makes continual monitoring and constant feedback mechanisms possible, it has been important in transforming performance management (Daugherty and Wilson 2018). AI analytics help distinguish between those who are performing well and those who require support for development. Another significant contribution of AI technologies is the replacement of automated performance evaluation with innovative practices and conventional annual performance assessment procedures (Tong et al. 2021).

Engagement and Retention of Employees: AI is widely used to track and enhance staff engagement. AI technologies evaluate employee sentiment through assessments and feedback, enabling HR to intervene before problems worsen (Marler and Boudreau 2017). AI can forecast stress and disengagement, providing useful strategies for sustaining talent (Guenole and Feinzig 2020).

Predictive Analysis and Workforce Planning: Predictive insights from AI are essential for staff planning (Malik et al. 2023). AI assists in forecasting future hiring requirements and productivity trends by examining historical data and patterns (Cappelli et al. 2019). As a result, HR is better equipped to allocate resources and supervise talent in a well-thought-out manner.

Automating Routine Work: Working efficiency is increased by AI's capacity to automate repetitive administrative duties like payroll disbursement and attendance tracking. Human error is reduced via robotic process automation (RPA), freeing up HR personnel to concentrate on critical assignments (Frey and Osborne 2017). This lowers expenses and makes resource management feasible (Brynjolfsson and McAfee 2014).

CHALLENGES AND ETHICAL CONSIDERATION

Even if the hospitality industry has access to a wide range of AI techniques and tools, the main issue is the security and reliability of these techniques and tools, as well as the best potential results that can be achieved in an efficient manner (Jaipuriar et al. 2024). The use of AI and social robot-based systems must still be restricted to highly structured environments and entails some high expenses, which are frequently deemed insufficient by the travel industry: financial expenses associated with purchasing, setting up, maintaining, and updating the technology, as well as employing experts to oversee it and teach workers. However, there are also non-financial expenses, particularly those related to employee resistance (Ivanov 2019, Ivanov and Webster 2017), where workers might experience psychological stress and fear (Li et al. 2019) from these developments and might decide not to utilize technology. In addition to staff resistance, resistance from customers may result in latent costs (Ivanov 2019, Ivanov and Webster 2017). Customers demand minimal effort for understanding how to operate technological gadgets, and if it is viewed as being too challenging, they may quit. Therefore, even when technological innovation is embraced to enhance the consumer experience, this result cannot be assumed (Ciuchita et al. 2019).

Concerns about AI's effects on tourism have mostly focused on job displacement (Tussyadiah and Miller 2019), because workers no longer feel like they belong, in addition to losing their employment (Li et al. 2019). Since the First Industrial Revolution, machines have been replacing human labor in the workforce. But for a very long time, machines could only take the place of people in basic, regular jobs. The latest generation of technology, including service robots, has emerged as a result of the development of AI and AI-powered technologies, and they are now capable of competing with and replacing humans in nearly every task (Brynjolfsson and McAfee 2011). In the next ten years, robots may displace 25% of the hospitality workforce, classifying the use of robots as an evolutionary transformational change. The reception area of hotels is one example of a typical tourism function that may even go extinct (Bowen and Morosan 2018).

Although there is an increase in efficiency, ethical issues are crucial to the use of AI. If AI receives instructions on biased data, it may reinforce biases in recruiting and performance reviews. AI's application in employee surveillance raises privacy concerns and may erode employee-employer trust. For AI to be used responsibly in HR, accountability and transparency are crucial. Data privacy is a major implementation difficulty for AI. Because AI depends on gathering a lot of data, security and permission from users are issues (Guan et al. 2025). One of the main obstacles to the use of big data in tourism is privacy concerns. According to the study, decision-makers in the Egyptian tourism authority frequently lack access to or control over consumer data, which hinders their capacity to successfully apply AI (Elsawy and Eltayeb 2023). Regulatory obstacles and security worries are frequently cited as reasons for organizational resistance to AI adoption (Elsawy 2023). This confirms earlier research showing that customer confidence depends on lucidity in AI data gathering and use (Buhalis and Sinarta 2019).

Excessive dependence on AI may decrease overall consumer pleasure by making hospitality encounters seem impersonal (Saleh 2025, Tussyadiah 2020). For example, when guests have complicated requests or grievances that AI systems may not be able to resolve, they frequently prefer human connection. Some jobs will be eliminated or drastically reduced in number due to automation, while other jobs will see changes in their duties and nature, and new employment positions will be created.

FUTURE TRENDS IN AI AND HOTEL EMPLOYEE PERFORMANCE

The tourist sector is seeing a rise in the importance of AI-driven efforts to promote sustainability, which provide ways to improve productivity and lessen environmental impact. According to Farinha et al. (2025), AI-driven energy management programs can drastically reduce hotel energy usage, which is consistent with broader sustainability objectives. Elsawy (2023) talks about how AI, as well as the adoption of e-business, helps travel businesses operate more efficiently and sustainably. Nevertheless, many businesses—particularly in developing nations—face major adoption challenges, such as high expenses and inadequate technological competence, even when they acknowledge the advantages of AI. According to Elsawy (2023), just 37.4% of Egyptian travel businesses utilize e-business, highlighting the major obstacles including high expenses and an abundance of infrastructure and expertise in technology. To fully realize AI's promise, personnel training and investments in technological infrastructure are required. The use of AI in the hospitality industry will keep rising, with a greater focus on smooth human-AI interaction. Long-term success will be attained by companies that successfully strike a balance between human-centered services and technology improvements ((Buhalis and Sinarta 2019, Mousa et al. 2025). AI will be essential for forecasting consumer behaviour, refining pricing schemes, and enhancing operational sustainability in hotels (Farinha et al. 2025, Tussyadiah 2020). In order to overcome obstacles to adoption and ensure that AI fosters long-term economic and environmental sustainability in addition to operational effectiveness, hospitality businesses need to focus on strategic expenditures and policy backing as AI-powered sustainability programs gain traction.

From the standpoint of the consumer, AI will enable them to plan their vacations faster, with much cheaper transaction costs, along with a completely customized package that fits their requirements and preferences. Cultural and linguistic disparities will be an added draw rather than a hindrance to travel. Customers will be able to get the greatest service possible thanks to technologies that protect their privacy to the greatest extent feasible (Bowen and Morosan 2018).

However, there is also a less optimistic viewpoint. According to this theory, automation will replace the majority of jobs in the sector, eliminating the sense of hospitality. As is now the case for numerous airports, there will be situations where employing machines is required rather than only a choice. Visitors will have to cope with robots and machinery that aren't prepared for use in a hospitality setting, and companies will elect to utilize less expensive goods even if they don't work as well (Bulchand-Gidumal 2022).

CONCLUSION

Employee performance could be greatly enhanced by integrating AI within the systems. Automating repeated tasks increases staff productivity and facilitates well-informed decision-making. Numerous factors determine how much AI can improve employee performance. These elements mostly include the willingness of employees to integrate and utilize AI channels, as well as managerial assistance, guidelines, and laws. The application of AI-driven HR solutions in the development and execution of HR processes and policies can help improve employee retention and participation. Additionally, by using AI tools, staff well-being and health can be efficiently assessed and tracked, which will increase staff motivation, engagement, and satisfaction. Employee concerns about invasion of privacy and substitution are also highlighted by the study. If organizations concentrate on creating a strategic and ethical structure to help and guide their staff, these issues can be appropriately addressed. The integration and cooperation of artificial intelligence and machines enhances innovation and operational efficiency, which in turn raises staff productivity. Strong management support makes it easier to align company objectives and goals with the customization of AI technologies, which raises staff productivity altogether.

Finally, the presence of the necessary infrastructure is the most significant and crucial element in the adoption and integration of AI systems. Adoption and integration go more smoothly for organizations with a contemporary, well-equipped digital infrastructure. By implementing and constructing an infrastructure that encourages AI integration and acceptance, management and staff will be able to obtain timely and useful insights, which will boost operational efficiency and ultimately boost employee productivity. Among the clear elements from the investigation that point to the development of contemporary workplaces are performance, involvement, and ethical considerations. However, future study on AI use will be crucial, particularly in the areas of biases and fairness.

The study's conclusion highlights the growing significance of AI in HRM and provides compelling justification for more investigation. The study could be enhanced by filling in the gaps of the research technique, incorporating research from various countries, and using AI ethically, which will be crucial for transforming the dynamics of productivity along with the nature of labor in the future.

REFERENCES

Aijaz Ali Khoso, S.H. 2025. Evaluating the Effectiveness of AI-Driven Personalized Recommendations in Enhancing Guest Experiences in Jeddah's Restaurants. International Research Journal of Management and Social Sciences. 6(2): 23–46. Retrieved from www.irjmss.com

Al-Hubaishi, M., Ali, M.A. and Tahir, N.M. 2024. Personalized Food Recommendations: A Machine Learning Model for Enhanced Dining Choices. 2024 IEEE 14th International Conference on Control System, Computing and Engineering (ICCSCE). 316–320. IEEE. doi:10.1109/ICCSCE61582.2024.10696536

Arney, K. 2017 'Algorithm's gonna get you', The Times Educational Supplement. Available at: https://www.tes.com/magazine/article/algorithms-gonna-get-you.

Baard, P.P., Deci, E.L. and Ryan, R.M. 2004. Intrinsic need satisfaction: A motivational basis of performance and well-being in two work settings. Journal of Applied Social Psychology. 34(10): 2045–2068. https://doi.org/10.1111/j.1559-1816.2004.tb02690.x

Batabyal, D. and Das, D.K. 2013. Management of travel booking in the light of destination marketing: Evidence from an Alpine Indian State-Sikkim. Indian Journal of Marketing. 43(11): 36–45.

Bateman, T.S. and Crant, J.M. 1993. The proactive component of organizational behavior: A measure and correlates. Journal of Organizational Behavior. 14(2): 103–118. https://doi.org/10.1002/job.4030140202

Belk, R.W., Belanche, D. and Flavián, C. 2023. Key concepts in artificial intelligence and technologies 4.0 in services. Service Business. 17(1): 1.

Bell, B.S. and Kozlowski, S.W. 2008. Active learning: Effects of core training design elements on self-regulatory processes, learning, and adaptability. Journal of Applied Psychology. 93(2): 296–316. https://doi.org/10.1037/0021-9010.93.2.296

Benko, A. and Lányi, C.S. 2009. History of Artificial Intelligence. *In*: Khosrow-Pour, M. (ed.). Encyclopedia of Information Science and Technology, 2nd Ed. (p. 4). IGI Global Scientific Publishing. doi:10.4018/978-1-60566-026-4.ch276

Bernardini, S., Fagnani, F. and Smith, D.E. 2018. Extracting mutual exclusion invariants from lifted temporal planning domains. Artificial Intelligence. 258: 1–65. doi:10.1016/j.artint.2018.01.004

Bersin, J. 2020. The rise of AI in HR: How artificial intelligence is transforming work. Deloitte Insights.

Bertsimas, D. and Tsitsiklis, J.N. 1997. Introduction to Linear Optimization. Athenas Scientific and Dynamic Ideas, Belmont. 608pp.

Bock, D.E., Wolter, J. and Ferrell, O.C. 2020. Artificial intelligence: Disrupting what we know about services. Journal of Services Marketing. 34(3): 317–334. https://doi.org/10.1108/JSM-01-2019-0047.

Borges, A.F., Laurindo, F.J., Spínola, M.M., Gonçalves, R.F. and Mattos, C.A. 2021. The strategic use of artificial intelligence in the digital era: Systematic literature review and future research directions. International Journal of Information Management. 57: 1–16. https://doi.org/10.1016/j.ijinfomgt.2020.102225

Boudreau, J.H. 2017. An evidence-based review of HR analytics. International Journal of Human Resource Management. 28(1): 3–26. doi:10.1080/09585192.2016.1244699

Bowen, J. and Morosan, C. 2018. Beware hospitality industry: the robots are coming. Worldwide Hospitality and Tourism Themes. 10(3): 726–733. doi:10.1108/WHATT-07-2018-0045

Brougham, D. and Haar, J. 2020. Technological disruption and employment: The influence on job insecurity and turnover intentions: A multi-country study. Technological Forecasting and Social Change. 161: 120276. https://doi.org/10.1016/j.techfore.2020.120276

Brynjolfsson, E. and McAfee, A. 2011. Race against the machine: how the digital revolution is accelerating innovation, driving productivity, and irreversibly transforming employment and the economy. Digital Frontier Press.

Brynjolfsson, E. and McAfee, A. 2014. The Second Machine Age: Work, Progress, and Prosperity in a Time of Brilliant Technologies. W.W. Norton & Company.

Bughin, J., LaBerge, L. and Mellbye, A. 2017. The case for digital reinvention. McKinsey Quarterly. 2017: 26–41.

Buhalis, D. and Sinarta, Y. 2019. Real-time co-creation and nowness service: Lessons from tourism and hospitality. Journal of Travel & Tourism Marketing. 36: 563–582. https://doi.org/10.1080/10548408.2019.1592059

Buhalis, D., Harwood, T., Bogicevic, V., Viglia, G., Beldona, S. and Hofacker, C. 2019. Technological disruptions in services: lessons from tourism and hospitality. Journal of Service Management. 30(4): 484–506. doi:10.1108/JOSM-12-2018-0398

Bulchand-Giduma, J. 2022. Impact of artificial intelligence in travel, tourism, and hospitality. *In*: Xiang, Z., Fuchs, M., Gretzel, U. and Höpken W. (eds). Handbook of e-Tourism, pp. 1943–1962. Springer. doi:10.1007/978-3-030-48652-5_110

Cappelli, P., Tambe, P. and Yakubovich, V. 2019. Artificial intelligence in human resources management: Challenges and a path forward. Academy of Management Perspectives. 33(2): 153–170. doi: 10.5465/amp.2017.0083

Carvalho, M., Kastenholz, E. and Carneiro, M.J. 2021. Pairing co-creation with food and wine experiences—A holistic perspective of tourist experiences in Dão, a Portuguese wine region. Sustainability. 13(23): 13416. https://doi.org/10.3390/su132313416

Castelvecchi, D. 2016. Can we open the black box of AI? Nature. 538(7623): 20–23. doi:10.1038/538020a

Chamorro-Premuzic, T., Akhtar, R., Winsborough, D. and Sherman, R. 2018. The Talent Delusion: Why Data, not Intuition, is Key to Unlocking Human Potential. Harvard Business Review Press.

Chen, Z.-B. and Liu, Y. 2020. Application of face recognition in smart hotels. *In*: 2020 IEEE Eurasia Conference on IOT, Communication and Engineering (ECICE), Yunlin, Taiwan, pp. 180–182. doi: 10.1109/ECICE50847.2020.9302014.

Choi, Y., Choi, M., Oh, M. and Kim, S. 2020. Service robots in hotels: understanding the service quality perceptions of human-robot interaction. Journal of Hospitality Marketing and Management. 29(6): 613–635. https://doi.org/10.1080/19368623.2020.1703871

Chowdhury, S., Budhwar, P., Dey, P.K., Joel-Edgar, S. and Abadie, A. 2022. AI-employee collaboration and business performance: Integrating knowledge-based view, socio-technical systems and organisational socialisation framework. Journal of Business Research. 144: 31–49. https://doi.org/ 10.1016/j.jbusres.2022.01.069

Chowdhury, S., Dey, P., Joel-Edgar, S., Bhattacharya, S., Rodriguez-Espindola, O., Abadie, A., et al. 2023. Unlocking the value of artificial intelligence in human resource management through AI capability framework. Human Resource Management Review. 33(1): 100899. https:// doi.org/10.1016/j.hrmr.2022.100899

Ciuchita, R., Mahr, D. and Odekerken-Schröder, G. 2019. Deal with it: How coping with e-service innovation affects the customer experience. Journal of Business Research. 103: 130–141. doi:10.1016/j.jbusres.2019.05.036.

Claveria, O., Monte, E. and Torra, S. 2015. A new forecasting approach for the hospitality industry. International Journal of Contemporary Hospitality Management. 27(7): 1520–1538. doi: https://doi.org/10.1108/IJCHM-06-2014-0286

Crotts, J., Mason, P. and Davis, B. 2009. Measuring guest satisfaction and competitive position: Application of stance analysis to blog narratives. Journal of Travel Research. 48(2): 139–151.

Daugherty, P.R. and Wilson, H.J. 2018. Human + Machine: Reimagining Work in the age of AI. Harvard Business Review Press.

Dirican, C. 2015. The impacts of robotics, artificial intelligence on business and economics. Proredia-Sorial arvi Behavioral Sciences. 195: 564–573. doi: 10.1016/j.sbspro.2015.06.134

Dixon, M., Ponomareff, L., Turner, S. and DeLisi, R. 2017. Kick-Ass Customer Service. Harvard Business Review. Jan–Feb 2017 Issue. Available from: https://hbr.org/2017/01/kick-ass-customer-service.

Doborjeh, Z., Hemmington, N., Doborjeh, M. and Kasabov, N. 2022. Artificial intelligence: A systematic review of methods and applications in hospitality and tourism. International Journal of Contemporary Hospitality Management. 34(3): 1154–1176. doi: 10.1108/ IJCHM-06-2021-0767

Dušan Mandić, A.P.-Š. 2023. Automated service delivery in hotels: Balancing efficiency and human interaction for optimal guest satisfaction. 1st International Scientific Conference on Economy, Management and Information Technologies-ICEMIT 2023. Belgrade, Serbia. doi: 10.46793/ICEMIT23.137M

Dwivedi, Y.K., Ismagilova, E., Hughes, D.L., Carlson, J., Filieri, R., Jacobson, J., et al. 2021. Setting the future of digital and social media marketing research: Perspectives and research propositions. International Journal of Information Management. 59: 102168. doi:10.1016/j.ijinfomgt.2020.102168

Elsawy, T.M. 2023. Determinants of e-business usage by travel agencies in developing countries: a decision tree approach. International Journal of Tourism Policy. 13(1). doi: 10.1504/ijtp.2023.129173

Elsawy, T.M. and Eltayeb, N. 2023. Examining the Adoption Barriers of Big Data Applications from the Perspective of Tourism Authorities in Egypt. Journal of the Faculty of Tourism and Hotels, Mansoura University. 13(13): 81–124. doi: 10.21608/mkaf.2023.299783

Elsawy, M.N. 2025. A Research report the role of artificial intelligence in transforming the hospitality and tourism industry. Pharos International Journal of Tourism and Hospitality. 4(1): 62–66. Retrieved from https://pijth.journals.ekb.eg/

Erica Mingotto, F.M. 2021. Challenges in re-designing operations and jobs to embody AI and robotics in services. Findings from a case in the hospitality industry. 31: 493–510. doi: 10.1007/s12525-020-00439-y

Farinha, C., Avelar, S. and Borges-Tiago, T. 2025. Service innovation snapshot from the lens of a top journal. *In*: Kavoura, A., Briciu, VA. and Briciu, A. (eds). Strategic Innovative Marketing and Tourism. ICSIMAT 2024. Springer Proceedings in Business and Economics. Springer, Cham. https://doi.org/10.1007/978-3-031-81962-9_67

Feinzig, N.G. 2020. The power of people: How AI is shaping the future of employee engagement and retention. IBM Institute for Business Value.

Frey, C.B. and Osborne, M.A. 2017. The Future of Employment: How Susceptible are Jobs to Computerization? Oxford University Press.

Fusté-Forné F. and Jamal, T. 2021. Co-Creating new directions for service robots in hospitality and tourism. Tourism and Hospitality. 2(1): 43–61. doi: 10.3390/tourhosp 2010003.

Gao, J. 2024. Optimizing hotel revenue management through dynamic pricing algorithms and data analysis. Journal of Computational Methods in Sciences and Engineering. 25(2). doi: 10.1177/14727978241298467

Gillet, N., Gagné, M., Sauvagère, S. and Fouquereau, E. 2013. The role of supervisor autonomy support, organizational support, and autonomous and controlled motivation in predicting employees' satisfaction and turnover intentions. European Journal of Work and Organizational Psychology. 22(4): 450–460. https://doi.org/10.1080/1359432X.2012.665228

Glikson, E. and Woolley, A.W. 2020. Human trust in artificial intelligence: Review of empirical research. Academy of Management Annals. 14(2): 627–660. https://doi.org/10.5465/annals.2018.0057

Guan, H., Yu, L., Zhou, L., Xiong, L., Chowdhury, K., Xie, L., et al. 2025. Privacy and Accuracy-Aware AI/ML model deduplication. *In*: Proceedings of the ACM on Management of Data, 3, pp. 1–28. doi:10.1145/3725340

Guenole, N. and Feinzig, S. 2020. The Power of People: How AI is Shaping the Future of Employee Engagement and Retention. IBM Institute for Business Value.

Gursoy, D.C. 2019. Consumers acceptance of artificially intelligent (AI) device use in service delivery. International Journal of Information Management. 49: 157–169. doi:10.1016/j.ijinfomgt.2019.03.008

Haefner, N., Wincent, J., Parida, V. and Gassmann, O. 2021. Artificial intelligence and innovation management: A review, framework, and research agenda. Technological Forecasting and Social Change. 162: 1–10. https://doi.org/10.1016/j.techfore.2020.120392

Haenlein, M., Kaplan, A., Tan, C.W. and Zhang, P. 2019. Artificial Intelligence (AI) and Management Analytics. Journal of Management Analytics. 6(4): 341– 343. https://doi.org/10.1080/23270012.2019.1699876

Hardt, S.B. 2019. Fairness in machine learning. Fairness, Accountability, and Transparency. doi: 10.1145/3287560.3287583

Hintze, A. 2016. Understanding the four types of AI, from reactive robots to self-aware beings. The Conversation. Retrieved from https://theconversation.com/understanding-the-four-types-of-ai-from-reactive-robots-to-self-aware-beings-67616

Iftikhar, P., Kuijpers, M.V., Khayyat, A., Iftikhar, A., DeGouvia De Sa, M. 2020. Artificial intelligence: A new paradigm in obstetrics and gynecology research and clinical practice. Cureus. 12(2): e7124. doi: 10.7759/cureus.7124.

Ivanov S.H. and Webster, C. 2017. Adoption of robots, artificial intelligence and service automation by travel, tourism and hospitality companies—A cost-benefit analysis. Prepared for the International Scientific Conference “Contemporary Tourism—Traditions and Innovations”, Sofia University, 19–21 October 2017. Available at SSRN: https://ssrn.com/abstract=3007577.

Ivanov, S.H. 2019. Ultimate transformation: How will automation technologies disrupt the travel, tourism and hospitality industries? Zeitschrift Für Tourismuswissenschaft. 11(1): 25–43. doi:10.1515/tw-2019-0003

Jaipuriar, S., Pal, S. and Rawal, Y.S. 2024. Transforming hospitality experiences: An exploratory study of service robots in the hospitality industry. *In*: AI Innovations for Travel and Tourism, pp. 121–134. IGI Global Scientific Publishing.

Javaid, Md., Haleem, A. and Singh, R.P. 2023. A study on ChatGPT for Industry 4.0: Background, potentials, challenges, and eventualities. Journal of Economy Technology. 1: 127–143.

Jiang, Z. 2017. Proactive personality and career adaptability: The role of thriving at work. Journal of Vocational Behavior. 98: 85–97. https://doi.org/10.1016/j.jvb.2016.10.003

Jundt, D.K., Shoss, M.K. and Huang, J.L. 2015. Individual adaptive performance in organizations: A review. Journal of Organizational Behavior. 36(S1): S53–S71. doi:10.1002/job.1955

Knowles, R.T. 2018. Teaching who you are: Connecting teachers’ civic education ideology to instructional strategies. Theory and Research in Social Education. 46(1): 68–109. doi:10.1080/00933104.2017.1356776

Kolasani, S. 2024. Revolutionizing manufacturing, making it more efficient, flexible, and intelligent with Industry 4.0 innovations. International Journal of Sustainable Development Through AI, ML and IoT. 3(1): 1–17.

Kong, H., Yuan, Y., Baruch, Y., Bu, N., Jiang, X. and Wang, K. 2021. Influences of artificial intelligence (AI) awareness on career competency and job burnout. International Journal of Contemporary Hospitality Management. 33(2): 717–734. https://doi.org/10.1108/IJCHM-07-2020-0789

Kurzweil, R. 2005. The Singularity is Near: When Humans Transcend Biology. New York: Penguin. Retrieved from https://paisdospuntocero.wordpress.com/wp-content/uploads/2018/04/book-kurzweil-singularity-is-near-1.pdf

Lai, W.C. and Hung, W.H. 2018. A framework of cloud and AI based intelligent hotel. *In*: Proceedings of the 18th International Conference on Electronic Business (pp. 36–43). ICEB, Guilin, China.

Law, R. 1998. Room occupancy rate forecasting: A neural network approach. International Journal of Contemporary Hospitality Management. 10(6): 234–239.

Levinthal, D.A. and March, J.G. 1993. The myopia of learning. Strategic Management Journal. 14(S2): 95–112. https://doi.org/10.1002/smj.4250141009

Li, J.J., Bonn, M.A. and Ye, B.H. 2019. Hotel employee's artificial intelligence and robotics awareness and its impact on turnover intention: The moderating roles of perceived organizational support and competitive psychological climate. Tourism Management. 73: 172–181. https://doi. org/10.1016/j.tourman.2019.02.006

Li, J. and Huang, J.S. 2020. Dimensions of artificial intelligence anxiety based on the integrated fear acquisition theory. Technology in Society. 63: 101410. https://doi.org/10.1016/j.techsoc.2020.101410

Li, M., Yin, D., Qiu, H. and Bai, B. 2021. A systematic review of AI technology-based service encounters: Implications for hospitality and tourism operations. International Journal of Hospitality Management. 95: 1–10. https://doi.org/10.1016/j.ijhm.2021.102930

Liang, X., Guo, G., Shu, L., Gong, Q. and Luo, P. 2022. Investigating the double-edged sword effect of AI awareness on employee's service innovative behavior. Tourism Management. 92: 1–14. https:// doi.org/10.1016/j.tourman.2022.104564

Liu, S.Q., Vakeel, K.A., Smith, N.A., Alavipour, R.S., Wei, C. and Wirtz, J. 2024. AI concierge in the customer journey: What is it and how can it add value to the customer? Journal of Service Management. 35(6): 136–158. doi: https://doi.org/10.1108/JOSM-12-2023-0523

Malik, N., Tripathi, S.N., Kar, A.K. and Gupta, S. 2021. Impact of artificial intelligence on employees working in industry 4.0 led organizations. International Journal of Manpower. 43(2): 334–354. https://doi.org/10.1108/IJM-03-2021-0173

Malik, A., Budhwar, P. and Kazmi, B.A. 2023. Artificial intelligence (AI)-assisted HRM: Towards an extended strategic framework. Human Resource Management Review. 33(1): 100940. https://doi.org/10.1016/j.hrmr.2022.100940

Mandić, D., Panić, A. and Čičin-Šain, M. 2023. Automated service delivery in hotels: balancing efficiency and human interaction for optimal guest satisfaction. International Scientific Conference on Economics, Management and Information Technologies. 1(1): 167–174. https://doi.org/10.46793/ICEMIT23.137M

Makarius, E.E., Mukherjee, D., Fox, J.D. and Fox, A.K. 2020. Rising with the machines: A sociotechnical framework for bringing artificial intelligence into the organization. Journal of Business Research. 120: 262–273. https://doi.org/10.1016/j.jbusres.2020.07.045

Markula, A. 2023. The Use of Artificial Intelligence in Dynamic Pricing Strategies. Aaltodoc. Information and Service Management. Retrieved from https://aaltodoc.aalto.fi/server/api/core/bitstreams/cbb8bc9a-5530-40f1-87c8-220410f77724/content

Marler, J.H. and Boudreau, J.W. 2017 An evidence-based review of HR analytics. The International Journal of Human Resource Management. 28(1): 3–26. https://doi.org/10.1080/09585192.2016.1244699

McCarthy, J., Minsky, M.L., Rochester, N. and Shannon, C.E. 1956. A Proposal for the Dartmouth Summer Research Project on Artificial Intelligence. doi:10.1609/aimag.v27i4.1904

Megalingam, R.K., Vadivel, S.R.R., Kotaprolu, S.S., Nithul, B., Kumar, D.V. and Rudravaram, G. 2025. Cleaning robots: A review of sensor technologies and intelligent control strategies for cleaning. Journal of Field Robotics. 42: 2234–2259. https://doi.org/10.1002/rob.22515

Mousa, M., Arslan, A. and Lange, T. 2025. When extreme events become the norm: How do artisan entrepreneurs adapt identity? International Journal of Organizational Analysis. 33(12): 35–55. doi: https://doi.org/10.1108/IJOA-07-2024-4647

Öksüz, M., Bulut, Ç., Candemir, A. and Bozkurt, İ. 2025. Digital or physical satisfaction? The effects of consumers' digital intentions and physical experiences on revisiting the restaurants. International Journal of Hospitality Management. 125: 104011. doi: 10.1016/j.ijhm.2024.104011

Osborne, C.B. 2017. The Future of Employment: How Susceptible are Jobs to Computerization? Oxford University Press.

Parker, S.K. and Grote, G. 2022. Automation, algorithms, and beyond: Why work design matters more than ever in a digital world. Applied Psychology. 71(4): 1171–1204. https://doi.org/10.1111/apps.12241

Pattie, D.C. and Snyder, J. 1996. Using a neural network to forecast visitor behavior. Annals of Tourism Research. 23(1): 151–164.

Poole, D.L. and Mackworth, A.K. 2010. Artificial Intelligence: Foundations of Computational Agents. Cambridge University Press. doi:10.1017/CBO9780511794797

PWC. 2020. Retrieved from PwC's Global Artificial Intelligence Study: Exploiting the AI Revolution: https://www.pwc.com/gx/en/issues/artificial-intelligence/publications/artificial-intelligence-study.html

Rane, N. 2023. Enhancing customer loyalty through Artificial Intelligence (AI), Internet of Things (IoT), and Big Data technologies: improving customer satisfaction, engagement, relationship, and experience. Internet of Things (IoT), and Big Data Technologies: Improving Customer Satisfaction, Engagement, Relationship, and Experience. doi: 10.2139/ssrn.4616051

Rasool, N. and Bhat, J.I. 2025. Artificial intelligence in restaurant. *In*: pp. 171–205. Zahoor, I., Ahmad Wani, S. and Ganaie, T.A. (eds). Artificial Intelligence in the Food Industry: Enhancing Quality and Safety. CRC Press. doi: 10.1201/9781032633602

Russell, S., 2022. Human-Compatible Artificial Intelligence. *In*: Muggleton, S. and Chater, N. (eds). Human-Like Machine Intelligence. 3–22. (Oxford, 2021; online edn, Oxford Academic, 19 Aug. 2021), https://doi.org/10.1093/oso/9780198862536.003.0001.

Russell, S.J. and Norvig, P. 2016. Artificial Intelligence: A Modern Approach. Hoboken, New Jersey, USA: Pearson.

Saleh, M.I. 2025. Generative artificial intelligence in hospitality and tourism: Future capabilities, AI prompts, and real-world applications. Journal of Hospitality Marketing and Management. 34(2): 189–208. doi: 10.1080/19368623.2025.2458603

Shah, T.R., Kautish, P. and Mehmood, K. 2023. Influence of robots service quality on customers' acceptance in restaurants. Asia Pacific Journal of Marketing and Logistics. 35(12): 3117–3137. https://doi.org/10.1108/APJML-09-2022-0780

Sharma, M., Tyagi, P.K., Semwal, R., Tyagi, P., Singh, V.J. and Kumari, N. 2024. Robotic concierges and smart spaces: A vision for the future of hospitality services. International Conference on Innovation and Regenerative Trends in Tourism and Hospitality Industry. 511–525. Atlantis Press. doi: 10.2991/978-94-6463-437-2_34

Shin, H.H. and Jeong, M. 2020. Guests' perceptions of robot concierge and their adoption intentions. International Journal of Contemporary Hospitality Management. 32(8): 2613–2633. doi: https://doi.org/10.1108/IJCHM-09-2019-0798.

Shirkhani, S., Mokayed, H., Saini, R. and Chai, H.Y. 2023. Study of AI-driven fashion recommender systems. SN Computer Science. 4(5): 514. doi: 10.1007/s42979-023-01932-9

Slemp, G.R., Kern, M.L., Patrick, K.J. and Ryan, R.M. 2018. Leader autonomy support in the workplace: A meta-analytic review. Motivation and Emotion. 42(5): 706–724. https://doi.org/10. 1007/s11031-018-9698-y

Soofastaei, A. 2021. Introductory Chapter: Virtual Assistants. London: IntechOpen: doi: 10.5772/intechopen.91579

Stank, T., Esper, T., Goldsby Thomas, J., Zinn, W. and Autry, C. 2019. Toward a digitally dominant paradigm for twenty-first century supply chain scholarship. International Journal of Physical Distribution and Logistics Management. 49(10): 956–971. https://doi.org/10.1108/IJPDLM-03-2019-0076

Sterne, J. 2017. Artificial Intelligence for Marketing: Practical Applications. Hoboken: Wiley. doi: 10.1002/9781119406341

Steven Van Belleghem. 2013. The Self Service Economy. Available from: https://www.stevenvanbelleghem.com/blog/new-report-the-self-service-economy/. (accessed 10.02. 2020)

Sumner, M. and Quinn, B. 2017. From concierge to superman: Perceptions of the contemporary hotel concierge in Edinburgh. International Journal of Culture, Tourism and Hospitality Research. 11(2): 243–254. https://doi.org/10.1108/IJCTHR-03-2016-0030

Tong, S., Jia, N., Luo, X. and Fang, Z. 2021. The Janus face of artificial intelligence feedback: Deployment versus disclosure effects on employee performance. Strategic Management Journal. 42(9): 1600–1631. https://doi.org/10.1002/smj.3322

Tussyadiah, I. and Miller, G. 2019. Perceived impacts of artificial intelligence and responses to positive behaviour change intervention. pp. 359–370. *In*: Pesonen, J. and Neidhardt, J. (eds). Information and Communication Technologies in Tourism 2019. Springer, Cham. https://doi.org/10.1007/978-3-030-05940-8_28.

Tussyadiah, I.P. 2020. A review of research into automation in tourism: Launching the Annals of Tourism Research Curated Collection on Artificial Intelligence and Robotics in Tourism. Annals of Tourism Research. 81: 102883. doi: 10.1016/j.annals.2020.102883

Ukpabi, D.C., Aslam, B. and Karjaluoto, H., 2019. Chatbot Adoption in Tourism Services: A Conceptual Exploration, *In*: Ivanov, S. and Webster, C. (eds). Robots, Artificial Intelligence, and Service Automation in Travel, Tourism and Hospitality. pp. 105–121. Emer-ald Publishing Limited. https://doi.org/10.1108/978-1-78756-687-320191006

Vrontis, D., Christofi, M., Pereira, V., Tarba, S., Makrides, A. and Trichina, E. 2022. Artificial intelligence, robotics, advanced technologies and human resource management: A systematic review. The International Journal of Human Resource Management. 33(6): 1237–1266. https:// doi.org/10.1080/09585192.2020.1871398

Wang, W., Chen, L., Xiong, M. and Wang, Y. 2021. Accelerating AI adoption with responsible AI signals and employee engagement mechanisms in health care. Information Systems Frontiers. 1–18. https://doi.org/10.1007/s10796-021-10154-4

Wang, K., Kong, H., Bu, X., Xiao, H., Qiu, N. and Li, J. 2022. AI in health tourism, developing a measurement scale. Asia Pacific Journal of Tourism Research. 27(9): 954–966. https://doi.org/10. 1080/10941665.2022.2142620

WEF. 2023. Future of Jobs Report 2023: Up to a Quarter of Jobs Expected to Change in Next Five Years. Retrieved May 21, 2023, from https://www.weforum.org/

Xun, Z. and Jiayun, Q. 2020. Research on hotel check-in management system based on deep belief network. Software Guide. 19(06): 125–129.

Index

W